ECOLOGY, ENGINEERING, AND MANAGEMENT

ECOLOGY, ENGINEERING, AND MANAGEMENT

Reconciling Ecosystem Rehabilitation
and Service Reliability

Michel J. G. van Eeten
Emery Roe

UNIVERSITY PRESS
2002

OXFORD
UNIVERSITY PRESS

Oxford   New York
Auckland   Bangkok   Buenos Aires   Cape Town   Chennai
Dar es Salaam   Delhi   Hong Kong   Istanbul   Karachi   Kolkata
Kuala Lumpur   Madrid   Melbourne   Mexico City   Mumbai   Nairobi
São Paulo   Shanghai   Singapore   Taipei   Tokyo   Toronto

and an associated company in Berlin

Copyright © 2002 by Oxford University Press, Inc.

Published by Oxford University Press, Inc.
198 Madison Avenue, New York, New York 10016

www.oup.com

Oxford is a registered trademark of Oxford University Press

All rights reserved. No part of this publication may be reproduced,
stored in a retrieval system, or transmitted, in any form or by any means,
electronic, mechanical, photocopying, recording, or otherwise,
without the prior permission of Oxford University Press.

Library of Congress Cataloging-in-Publication Data
Eeten, Michel van.
Ecology, engineering, and management : reconciling ecosystem
rehabilitation and service reliability / Michel van Eeten and Emery Roe.
p. cm.
Includes bibliographical references (p.   ).
ISBN 0-19-513968-2
1. Ecosystem management.   2. Environmental policy.   I. Roe, Emery.   II. Title.
QH75 .E34 2001
333.95—dc21   2001021711

9 8 7 6 5 4 3 2 1

Printed in the United States of America
on recycled, acid-free paper

We dedicate this book to Brian, Louise, Annemarie, and Scott

Contents

Acronyms ix

1 The Paradox of the Rising Demand for Both a Better Environment and More Reliable Services   3

2 The Paradox Introduced: Concepts and Cases   13

3 Adaptive Management in a High Reliability Context: Hard Problems, Partial Responses   51

4 Recasting the Paradox through a Framework of Ecosystem Management Regimes   85

5 Ecosystems in Zones of Conflict: Partial Responses as an Emerging Management Regime   129

6 Ecosystems in Zones of Conflict: The Case for Bandwidth Management   169

7 The Paradox Resolved: A Different Case Study and the Argument Summarized   217

Appendix: Modeling in the CALFED Program   241

Notes   245

References   251

Index   263

Acronyms

ACE	U.S. Army Corps of Engineers
AEA	Adaptive Environmental Assessment
ATLSS	Across Trophic Level System Simulation
BLM	U.S. Bureau of Land Management
BPA	Bonneville Power Administration (Columbia River Basin)
BR	U.S. Bureau of Reclamation
CDR	coupling–decoupling–recoupling (dynamic)
CMARP	Comprehensive Monitoring, Assessment and Research Program (Bay–Delta)
CRBSUM	Columbia River Basin Succession Model
DFG	California Department of Fish and Game
DWR	California Department of Water Resources
EAA	Everglades Agricultural Area
ECO-OPS	Ecosystems Operations Branch (proposed; Bay–Delta)
EDT	evaluation, diagnosis, and treatment (Columbia River Basin)
EIR	Environmental Impact Report
EIS	Environmental Impact Statement (Columbia River Basin)
EPA	U.S. Environmental Protection Agency
ERP	Ecosystem Restoration Program (Bay–Delta)
ESA	Endangered Species Act
EWA	Environmental Water Account (Bay–Delta)
FS	U.S. Forest Service
FWS	U.S. Fish and Wildlife Service
GIS	geographic information system
GPM	General Planning Model (Everglades)
HRO	high reliability organization

ICBEMP	Interior Columbia Basin Ecosystem Management Project
MAF	million acre feet
MWD	Metropolitan Water District (Southern California)
NEPA	National Environmental Protection Act
NMFS	U.S. National Marine Fisheries Service
NPPC	Northwest Power Planning Council
NPS	U.S. National Park Service
OCO	Operations Control Office (California Department of Water Resources)
PIT	Passive Integrated Transponder
ROD	Record of Decision
SFWMD	South Florida Water Management District
SIT	Science Integration Team
SWP	State Water Project (California)
SWRCB	State Water Resources Control Board
TNC	The Nature Conservancy
USDA	U.S. Department of Agriculture
USGS	U.S. Geological Survey
VAMP	Vernalis Adaptive Management Plan (Bay–Delta)
WCA	Water Conservation Area (Everglades)

ECOLOGY, ENGINEERING, AND MANAGEMENT

1 The Paradox of the Rising Demand for Both a Better Environment and More Reliable Services

The examples go on and on: loading fish in trucks and on barges to enable them to swim downstream; opening a water gate and drowning endangered birds in one area, or closing the gate and risk burning out habitat of the same species someplace else; spending more than $400 million a year to protect a handful of endangered species in just one region of a country; hatching endangered fish that end up too fat or stick out like neon in the water once released; releasing salmon trained to come to the surface for hatchery food when what is actually dropping from the sky are the ducks ready to eat them; keeping water in a reservoir to save the fish there, thus sacrificing other fish downstream; building a 250-foot-wide, 300-foot-high, $80 million device to better regulate the water temperature for salmon eggs in just one reservoir; controlled burning for fuel load management in the forests that harms not only air quality but also chronically bleeds pollution into adjacent aquatic ecosystems; breeding the wild properties out of endangered fish and releasing them, thereby polluting the gene pool of river fish; fighting urbanization to protect a green and open area, thereby condemning that area to monotonous, industrial agriculture and worse; closing a gate or releasing reservoir water in reaction to a sample of fish coming downstream and triggering electrical blackouts or the most severe urban water quality crisis in decades; restoring natural floodplains, erasing some of the oldest, best preserved, and greenest cultural landscapes in a country; putting in place even more massive infrastructure to keep ecosystems natural, thereby imprisoning them in intensive care units for life; and more.[1]

For some readers, these examples may appear a mix of the ridiculous and the desperate. Yet they are prime examples of a hard paradox at work: how do you reconcile the public's demand for a better environment which requires ecosystem improvements with their concurrent demand for reliable services from that environment, including clean air, water, and power? More formally, how do you meet this twofold management goal: (1) where decision makers are managing for reliable ecosystem services, they are also improving the associated ecological functions, and/or (2) where they are managing for improved ecological functions, they are better ensuring the reliability of ecosystem services associated with those functions? (For ease of exposition, we use "decision maker" broadly to cover any person or institution, public or private, making decisions to manage an ecosystem, its functions, or its services.) Less formally, how do we save the salmon and still have electricity? How do we save the delta smelt and still have urban water fit to drink? How can we keep an ecosystem natural when doing so makes it permanently dependent on "unnatural" intensive management by humans? How do we have a good environment and at the same time a prosperous economy? "Taking bold actions to preserve California's quality of life and environment in the face of a strong economy will make California even more attractive to workers and entrepreneurs. This is the paradox of a strong economy. Yet it is the only realistic chance for Californians to have both economic prosperity and a great place to live," according to one economist (Levy 1998, p. 5). The question, of course, is just how do we do this? There are answers to these questions. But they require a total rethinking of the paradox. As we will see, the answers pull decision makers to the ecosystem and landscape levels, where the crucial tradeoffs and priorities are best articulated and struck.

For some, there is no paradox. Just do not cut down the trees. Do not drive cars. Take out the dams. Let nature take care of itself. Or, from another perspective, "Do not screw around with the drinking water of millions of urban residents for the sake of one endangered species." People matter more than trees. Besides, economic growth is the best way to get people to want to save the environment. We do, however, believe hard issues are involved here that are not easily addressed. The buzz terms may come and go—"ecosystem management" or "industrial ecology" today, something different tomorrow—but the paradox is real and difficult to resolve.

Our book provides a conceptual framework, empirical case analyses, and an organizational proposal to resolve the paradox, be it in the United States, Europe, or wherever the hard issues are found. You can manage the seemingly unmanageable, reconcile the seemingly irreconcilable, make predictable the seemingly unpredictable, but it is clear that there are no easy answers here. In our search for answers we need a new framework, empirical analysis and management proposal that recasts the paradox in different policy terms. That framework must reconceive the real tradeoffs and priorities that confront ecologists, engineers, and others who take a better environment and service reliability seriously.

Thus, the book has multiple audiences. First are the key professions involved in the protection and improvement of ecosystems and in the provision and delivery of services from those ecosystems. These are ecologists (broadly writ to include other natural scientists such as conservation biologists, climatologists, forest scientists, and toxicologists); engineers (broadly writ to include hydrologists, environmental engineers, civil engineers, and for the time being, line operators); modeling and gaming experts; managers, planners, and policymakers; environmentalists and environmental groups; and other stakeholders such as water, power, agriculture, and recreation communities. Academic researchers, postdocs and graduate students in ecology, conservation biology, engineering, the policy sciences, and resource management, as well as those interested in interdisciplinary approaches in these fields will also find the book helpful. Finally, those interested in the Everglades, the Columbia River Basin, the San Francisco Bay–Delta, and the Green Heart of western Netherlands will find new insights here. As these cases are empirical examples of the same paradox faced elsewhere, we hope that our audiences are international throughout.

Although we take ecosystem management as our point of departure, the book is very much oriented to engineers in the natural resource fields. Ecosystem management has come about in response to the consequences of large-scale engineering of natural resources, such as major water and power generation works. Engineering and ecology thus have long been connected historically. Our approach is the first to integrate ecological and engineering considerations into a single ecosystem-based conceptual framework. In the process, we identify key professional blind spots and the opportunities they pose for meeting the twofold management goal. We show that even though engineers have been part of the problem of environmental degradation, they are also a key part of its solution. We do so by reinterpreting the engineering of reliable ecosystem services as a form of ecosystem management, which we call high reliability management.

Even disciplines at the interface of ecosystem services and functions, such as environmental engineering and restoration ecology, need a new approach to the paradox. Consider this exchange on the importance of ecological indicators in ecosystem management, which transpired while driving to a consultancy meeting in Sacramento:

Engineer: "... and if the indicator can't be measured, like connectivity of wetlands, we'll find a better one that can."
Ecologist: "But that's not the point."
Engineer: "Huh? We'll just come up with a better one, that's all."
Ecologist: "You can't. It doesn't matter if connectivity can't be measured. It's part of what makes a wetlands healthy. Take away connectivity and you won't have ecosystem integrity."

Thus, the crux of the matter is, what are the components without which we would have no ecosystem, let alone one in which decision makers can meet the twofold management goal? Clearly, the issue at hand is more than doing environmentally sensitive engineering or engineering-informed ecology. Ecologists, engineers, and other professions involved in meeting the twofold management goal currently see the goal in very different conceptual and practical terms. Each profession has its own culture, including their blind spots—or as one high reliability specialist terms them, "well-earned areas of ignorance" (T. R. La Porte 2000, pers. comm.). Our approach to managing ecosystems in zones of conflict between population growth, intensive resource utilization, and rising demand for a better environment makes the paradox more tractable to the various disciplines important for its resolution. The key, as the reader will see, is redefining ecosystem services and functions so that they can be both improved and recoupled.

The Book's Organization

Chapter 2, "The Paradox Introduced: Concepts and Cases," undertakes an up-to-date review of the literature on ecosystem management to provide a current conceptualization of ecosystem functions and services and their relation to each other. The chapter answers a series of questions, particularly: what is ecosystem management? Why ecosystem management? What are the threats to ecosystem management? What needs to be in place for ecosystem management? And who matters in ecosystem management? In answering the questions, the major paradox is explored more fully. The importance of adaptive management in ecosystem management as the main mechanism to address the paradox is drawn out and highlighted. Chapter 2 concludes with an introduction to the three U.S. case studies and ecosystem management initiatives in the San Francisco Bay–Delta, the Columbia River Basin, and the Everglades (the fourth case study is dealt with entirely in chapter 7).

Chapter 3, "Adaptive Management within a High Reliability Context: Hard Problems, Partial Responses," focuses on the many problems associated with undertaking adaptive management in political, cultural, and social situations that demand high reliability in service provision and delivery. The problems, which at their most obvious range from issues of research design through difficulties with trial-and-error learning to unforgiving politics, are discussed, as well as partial responses and remedies to address these problems. Building on our case studies, the chapter discusses in detail how the implementation of adaptive management specifically and ecosystem management generally have at a deeper level been hampered by increased human population growth and resource extraction, the variable availability of adequate conceptual and biological models (particularly for aquatic species and ecosystems), the many organizational goals and requirements that compete with achieving better management of ecosystems, and the rising demand for multiple services from ecosystems and their various resources. The persisting problems and unresolved topics call for a new approach to better address them.

Chapter 4, "Recasting the Paradox Through a Framework of Ecosystem Management Regimes," sets out our theoretical framework to address the twofold management goal. How do decision makers reintroduce to ecosystems some of the complexity and unpredictability that they once had without compromising the predictability of the services derived from these now-improved ecosystems?

Our answer revolves around a framework for distinguishing different ecosystems and their management regimes along a gradient with five dimensions. The dimensions, identified in chapter 3, are important in implementing adaptive management: low to high population densities, little to high resource extraction, few to many causally adequate models, competing considerations for ecosystem and organizational health, and few to many services expected from the ecosystem being considered. Four management regimes—self-sustaining management, adaptive management, case-by-case management, and high reliability management—are identified and discussed, from the "low" end to the "high" end of the five-dimensional gradient. These features are summarized in table 1.1.

A principal argument of chapter 4 is that, while adaptive management is appropriate for less populated and utilized ecosystems, many ecosystems fall in zones of conflict between population, resources, and environment, where management has to be more case-by-case and where the pressures of increasing population and extraction continually drive decision makers to seek high reliability management of ecosystem services. Using a new conceptual language and its implications (developed in chapters 5 and 6 as well), we propose that decision makers meet the management goal of improving services and functions simultaneously by undertaking a series

Table 1.1 Ecosystem Management Regimes Framework

	Corresponding Management Regimes			
Dimensions	Self-sustaining management	Adaptive management	Case-by-case management	High reliability management
Population densities	Low	←————————————→		High
Resource extraction	Little	←————————————→		High
Causally adequate models	Few	←————————————→		Many
Ecosystem and organizational health	Ecosystem health, with organizational implications	←————————————→		Organizational health, with ecosystem implications
Ecosystem (re)sources and services	Multiple sources/few services	←————————————→		Single resource/many services

of steps: matching ecosystem with management regime; redefining services and functions along the gradient; managing case-by-case when in zones of conflict; managing in zones of conflict through bandwidth management; managing in zones of conflict through settlement templates; networking the small and large scale; and funding and undertaking more science, engineering, adaptive management, and model-based gaming in ecosystem management initiatives.

Chapter 5, "Ecosystems in Zones of Conflict: Partial Responses as an Emerging Management Regime," takes a second look at the partial responses discussed in chapter 3. With the new framework in mind, we are now able to identify the partial responses not as unsuccessful attempts to do adaptive management, but in fact, as an emergent form of the case-by-case management regime in the Everglades, Columbia River, and Bay–Delta ecosystems. These areas are zones of conflict because increasing population, intensified resource utilization, and the rising demand for a better environment are clearly at odds with each other. The core challenge in these zones is to better recouple ecosystem functions to services. That means building a crosswalk between the regimes of adaptive management and high reliability management.

Here, we find an organizational mechanism at work that we call the coupling–decoupling–recoupling (CDR) dynamic. At the policy level, each of the cases exhibits a broad recognition of the need to couple functions and services. Each program is committed to an integral approach to water quality, water supply, protection of endangered species, ecosystem restoration, hydropower generation, and flood control. Yet attempts at policy coupling have generated their own organizational counterforces. Interlocked issues, as we will see, become unmanageable when different and already complex policies are correspondingly treated as just as tightly coupled as the world they seek to change. When faced with such a turbulent task environment, the pressure is to decouple issues of specific interest from that environment and buffer them in their own programs, agencies, or distinct professions. However, while the decoupling achieves short-term reductions in turbulence and increases program stability and effectiveness, it ends up undermining the very comprehensive approach that drove the initial systemwide coupling. Where the initial coupling generated pressure to decouple, the decoupling reinforces pressure to recouple at the operational level—the third element of the CDR dynamic. A crucial characteristic of operational recoupling, when it occurs, is that it enables a dynamic optimization process among different (often interagency) goals and issues, thereby rendering program and agency boundaries permeable. In fact, we find that decoupling is a necessary and positive part of the overall dynamic. Without it, recoupling would be impossible, though recoupling is never an automatic or linear next step. Using the CDR dynamic we revisit the organizational responses singled out in chapter 3 to identify innovations most effective in recoupling functions and services.

Chapter 6, "Ecosystems in Zones of Conflict: The Case for Bandwidth Management," starts by analyzing the cultures of the key professions involved in ecosystem management in zones of conflict: line operators, ecologists, engineers, species-specific regulators, modelers, and scientists. The analysis extends the CDR dynamic into professional culture, pointing out both blind spots and opportunities for the professions to cooperate more successfully in recoupling services and functions. Building on these opportunities and on the innovations discussed in chapter 5, we then develop the case for bandwidth management as our major organizational proposal.

Bandwidth management distinguishes two core processes: managing within the bandwidths and setting the bandwidths. The former consists of the operational recoupling of services and functions discussed earlier. However, some of the bandwidths set for line operators are conflicting and cannot be resolved by them. The conflict is then pushed into the process of setting the bandwidths. In a variety of ways, the discrepancies between bandwidths drive the agencies and professions involved to identify tradeoffs and set priorities. This also entails redrawing system boundaries and redefining the functions and services to be improved. We advance the notions of settlement templates, their networks, and a positive theory of the small scale to strengthen and improve this process of recoupling across different scales. The chapter concludes with concrete organizational proposals to improve bandwidth management in our most in-depth case study, the CALFED Program.

Chapter 7, "The Paradox Resolved: A Different Case Study, Wider Policy Implications, and the Argument Summarized," introduces our last case study, from the Netherlands, in order to draw out policy-oriented implications and recommendations of our management-oriented framework, case material, and organizational proposals. The chapter includes a synopsis of the book's arguments, themes, and recommendations.

Acknowledgments

Our list is long, but we must thank three people in particular: Kirk Jensen, our editor at Oxford University Press; Shiv Someshwar at the Rockefeller Foundation for supporting the initial research; and Peter Gratzinger, who coauthored the original report we did for the Foundation. We also thank the Rockefeller Foundation and Delft University of Technology for their financial assistance during the writing.

The people who reviewed the book draft in whole or in part are Louise Comfort, Robert Frosch, Janne Hukkinen, Garry Peterson, Todd La Porte, and Diane Rocheleau. They did their best and the errors that remain are ours. We had two sets of interviewees, one for the initial Rockefeller report and a slightly overlapping set for the book itself, all of whom we thank here: Stuart Appelbaum, Scott Bettin, Rob Cooke, Curtis Creel, Dick Daniel, Steven Davis, Don DeAngelis, Greg Delwiche, Susan Giannettino, Lance Gunderson, Tom Hagler, Bruce Herbold, John Hyde, Steve Johnson, Margaret Johnston, Peter Kiel, Wim Kimmerer, Gwen Knittweis, Pamela

Matson, Willis (Chip) McConnaha, Dennis O'Bryant, Ron Ott, Jatinder Punia, Thomas Quigley, Pete Rhoads, Curt Schmutte, Sally and Jim Shanks, Whendee Silver, Jim Spence, Tommie Strowd, Barbara Wales, Leo Winternitz, John Winther, and Rick Woodard. All interviewees were invited to review and amend their quotes in the text, which we edited beforehand only to reduce pauses, repetitions, and other infelicities of common speech. The only thing our interviewees are to be blamed for is the unexpected length of this book.

To maintain confidentiality, none of those interviewed is named in the text. We appreciate that this decision may be controversial. An important factor at play here is the difference in conventions between the natural and social sciences. In our experience, it is much more common to assume confidentiality and anonymity in interviewee comments when you are in the social sciences (both of us are from the policy analysis field) than in the natural sciences. Natural scientists believe that it is important to name who is saying what about a given issue related to science and research and give credit where credit is due. Social scientists, on the other hand, are very reluctant to "name names" of interviewees involved in highly polarized, acrimonious controversies, in the belief that to do so only personalizes the controversy all the more. We have sided with the social scientists in this book.

When discussing our initial research ideas, Steve Edwards suggested putting both adaptive and high reliability management within a broader management framework that treated them less as opposites than as management regimes triggered by different thresholds. An evening with Louise Fortmann and Donald Moore led to the first version of figure 4.1, which summarizes the main elements of our framework. Craig Thomas and Gene Bardach helped us track down the literature on interagency collaborations. At an earlier stage, Gene Rochlin helped us think through issues of high reliability. Comments from Randy Brown and Victor Pacheco on the CALFED Program were useful. Discussions at a conference of the Bay–Delta Modeling Forum with Russ Brown, Erwin Van Nieuwenhuyse, and Larry Smith were especially informative. Short conversations with Dave Fullerton, Evert Lindquist, Dave Sunding, Spreck Rosekrans, Terry Young, and David Zilberman were also instructive. We thank Ernest Alexander, Timothy Beatley, Laurens Jan Brinkhorst, Andreas Faludi, Bart Gremmen, Tjeerd de Groot, Maarten Hajer, Ernst ten Heuvelhof, and Kees de Ruiter, for helping us think through the Green Heart issues. Background information and editorial assistance has been provided by Michelle Luijben Marks, Wijnand Veeneman, Rob Schmidt, and Bart van Konijnenburg. As we wanted to stay close to the original materials, we have left the data quoted in this book a mixture of metric and imperial units. Darin Jensen was instrumental in producing the maps of the four areas of our case studies. Anonymous reviewers with *Environmental Management*, *Policy Sciences*, the *Journal of the American Planning Association*, and Oxford University Press deserve our final thanks.

Ideas developed in this book were first presented at seminars, panels, and workshops supported by Delft University of Technology; University of

California, Berkeley; the Netherlands' Ministry of Agriculture, Nature Management and Fisheries; the Netherlands' Institute for Inland Water Management and Waste Water Treatment (RIZA); the Association for Public Policy Analysis and Management, Wageningen University; and the Conference Group on Theory, Policy and Society. Portions of chapters 3 and 7 appear in initial versions in the following publications:

> Roe, E. M. 2001. Varieties of issue incompleteness and coordination. *Policy Sciences*, Vol. 34(2): 111–133. We kindly acknowledge permission to reproduce material given by Kluwer Academic Publishers.
>
> Roe, E. M. and M. van Eeten. 2001. Threshold-based resource management: a framework for comprehensive ecosystem management. *Environmental Management* Vol. 27(2): 195–214. We thank Springer-Verlag New York for permission to reproduce the material.
>
> Roe, E. M. and M. van Eeten. 2001. The heart of the matter: a radical proposal. *Journal of the American Planning Association* 67(1): 92–98.
>
> Roe, E. M., M. J. G. van Eeten, and P. Gratzinger. 1999. *Threshold-based resource management: the framework, case study and application, and their implications*. Report to the Rockefeller Foundation. Berkeley: University of California.
>
> Van Eeten, M. and E. M. Roe. 2000. Bouw huizen en kantoren helpt natuur. *Trouw* 22 July: 18.
>
> Van Eeten, M. and E. M. Roe. 2000. When fiction conveys truth and authority: The Netherlands Green Heart planning controversy. *Journal of the American Planning Association* 66(1): 58–76.

The conceptualizing and writing of the book has been a totally mutual and equal effort. Plus, it has been fun.

To end where this book really began, we dedicate it to Brian, Louise, Annemarie, and Scott.

2 The Paradox Introduced

Concepts and Cases

To recapitulate, the hard paradox is this: how do you improve ecological functions and related human services at the same time, if not everywhere then at least over the ecosystem and landscape as a whole? How do decision makers meet the twofold recoupling goal: (1) where they are managing for reliable ecosystem services, they would also be improving the associated ecosystem functions, and/or (2) where they are managing for improved ecosystem functions, they would also be better ensuring the reliability of the ecosystem services associated with those functions. In short, how do decision makers recouple ecosystem functions and services that over time have been decoupled to their detriment?

A set of terms have just been introduced that require explanation. The terms "recoupling," "decoupling," and, by implication, "coupling" are central to the arguments of our book and are formalized more fully in later chapters. (The controversial terms, "functions" and "services," are discussed in the next section.) Basically, the literature uses the former terms to refer to biophysical connections, organizational connections, or both. An example of the first is Ausubel (1996, pp. 1, 7, 8), who notes that agricultural modernization has meant "food decoupled from acreage" through the production of more crops on less land. Advances in science and technology "increasingly decouple our goods and services from the demands on planetary resources." Ausubel adds that we can expect "further decoupling [of] food from land. For more green occupations, today's farmers might become tomorrow's park rangers and ecosystem guardians. In any case, the rising yields, spatial contraction of agriculture, and sparing of land are a powerful antidote to the current losses of biodiversity and related environmental ills." Opschoor (1995) speaks of a similar technological phenomenon, "delinking," where rising incomes are decoupled over time from intensive material use. Also, the third Dutch national environmental policy plan seeks as one of its goals the decoupling of economic growth from environmental pollution (Ministry of Housing, Spatial Planning and Environment 1998).

These uses of "decoupling" all refer to the relation between services and environmental degradation. We, on the other hand, are talking about the relation between services and environmental assets, that is, ecosystem functions. Thus, another way to say "decoupling economic growth from air pollution" is to say "recoupling economic growth to improved air quality." Actually that is not completely true. As our book will show, reducing environmental degradation does not necessarily improve environmental assets. That is why the twofold goal speaks of the "recoupling" of improved functions and services.

Alongside these biophysical relations, the terms coupling, decoupling, and recoupling are used to describe organizational or policy connections, as we did ourselves in chapter 1. Goldstein (1999, pp. 248, 249), for instance, laments that current ecosystem management approaches and their advocates "decouple" wildlife and biodiversity preservation from conservation:

> While detaching species from management goals, some authors have attempted to justify paradigms with appeals to notions of habitat "quality" or "integrity," and the values of rare species are thus articulated as indicators of some attribute of natural areas rather than as elemental concerns in and of themselves. ... Ecosystem management ... has become associated with the decoupling of organismal information from conservation strategy.

Farrier (1995, p. 90, online version) describes the current proposals for "'decoupling' farmer income support from production of agricultural commodities, while 'recoupling' [support] to a 'green' commodity that is in increasingly short supply but that the market offers little incentive to produce." Sometimes it is unclear if the focus is on the organizational or the biophysical connections. For example, Barrett et al. (1999, p. 199), writing from an agroecoystem perspective, conclude "[i]t is now imperative to couple the heterotrophic urban environment with the autotrophic agricultural environment if societies are to establish or manage sustainable landscapes" (see Light et al. 1995 for a more explicit coupling between social and ecological systems).

Our use of the terms organizational coupling, decoupling, and recoupling is elaborated in chapters 4–6. There, the terms are used to explain the dynamics encountered in the ecosystem management initiatives of our case studies. For now, it is sufficient to recognize the biophysical and organizational dimensions of these concepts. In terms of the twofold management goal, each dimension defines recoupling differently, but with complementary effects. For the biophysical dimension, recoupling means improving the overall *set* of functions and services, rather than improving each function and service on its own, because the level of analysis is now the whole ecosystem or landscape. Improving the set means reducing the degree to which functions or services undermine each other. The easy answer, of course, is to insist that one take overall priority over the other: reduce services to restore functions (e.g., decrease population and extraction), or give priority to services (e.g., amend species-protection legislation to allow for more exceptions). Such proposals typically entail massive political, social, and cultural changes. Realistically, therefore, the pressure is always to find other ways to recouple functions and services.

Thus, the paradox remains. As there are many kinds of improvements to be sought and degradations to be avoided, how do decision makers decide which ones to give priority? The process of getting to a redefined set of functions and services is through organizational recoupling. For many people, the answer is bigger budgets, more coordination, and maybe even an overall agency with the authority to make tough decisions. Yet, as we will see, these answers have been tried without much success. For us, the more realistic option is to rethink what these budgets, coordination, and authority are about. From the perspective of the twofold management goal, organizational recoupling means the dynamic optimization of ecosystem functions and services. That is, striking tradeoffs and setting priorities within the overall set. The notion of organizational recoupling along with dynamic optimization is developed at length in the later chapters. Here, we focus, as does the literature on ecosystem management, on the biophysical dimension of recoupling.

There are two reasons for focusing on the ecosystem management literature rather than that on engineering, even though both address biophysical relationships. First, the former deals with both functions and services more directly than the latter. Second, we address engineering issues directly when we reconceive certain forms of ecosystem management as the high reliability management of natural resources. Our literature review proceeds through a series of questions about ecosystem management and the answers given to them.

What Are Ecosystem Functions and Services?

There are many ecosystem functions: regulating atmospheric chemical composition, temperatures, and precipitation; decomposing compounds; producing biomass; maintaining balances in carbon dioxide and nitrogen; permitting recovery from natural disturbances; filtering ultraviolet radiation; and cycling nutrients; among others. These functions in turn yield many potential benefits—hereafter termed ecosystem services—including commodities (such as timber, fish, and wildlife), specific services (such as hydropower, biological control, or pollution abatement), intangibles (such as preservation of open landscapes, endangered species, and wilderness), and amenities (such as places for recreation) (O'Neill et al. 1996, p. 23; Lackey 1998, p. 23; Strange et al. 1999, table 1).

The distinction between functions and services (or their cognates) is controversial (S. A. Levin 2000, pers. comm.). First, use of the terms is inconsistent in the literature. At times, services are distinguished from amenities, commodities, or goods and intangibles or values. Second, there are times when it is difficult to distinguish what are functions and what are services, for example, when the function of regulating the climate is coupled with the service of climate regulation (see table 2.1). Indeed, functions have been explicitly equated with services (e.g., Callicott et al. 1999, p. 27). Third, not only are functions difficult to distinguish from ecosystem processes, but there are commentators who also insist that the terms, "functions" and "processes," are at best ill-defined buzzwords with little empirical content on their own (e.g., Goldstein 1999). That said, we follow Costanza et al. (1997) and much of the literature in distinguishing between functions and services as in table 2.1, recognizing, of course, that in the field, case by case, the two may be quite closely related.

Some services, for example, food production, are in fact the benefit of several functions together. Similarly, since functions and services may be so closely coupled as to be difficult to distinguish on the ground, their management at the ecosystem level becomes both an opportunity and a challenge.

What Exactly Is Ecosystem Management?

That is just the problem. Apart from a focus on functions and services together, little is exact about ecosystem management. Consider some fairly representative summaries of the field:

> Ecosystem management has been defined in a variety of ways. In general, however, there is agreement that a goal of ecosystem management is to sustain ecosystem health, integrity, diversity and resilience to disturbances. ... This is achieved through the maintenance of productivity, biodiversity, landscape patterns, and an array of ecological functions and processes. ... It also requires the integration of social, economic, and ecological considerations at broad spatial and temporal scales (Cortner et al. 1998, p. 160).

> The diversity of definitions provides some indication of the current amorphous nature of the concept. ... Typical of definitions of ecosystem management are: (1) "A strategy or plan to manage ecosystems to provide for all associated organisms, as opposed to a strategy or plan for managing individual species." ... (2) "The careful and skillful use of ecological, economic, social, and managerial principles in managing ecosystems to produce, restore, or sustain ecosystem integrity and desired conditions, uses, products, values, and services over the long-term." ... (3) "To restore and maintain the health, sustainability, and biological diversity of ecosystems while supporting sustainable economies and communities" ... [My own] definition of ecosystem management is:

Table 2.1 Partial List of Ecosystem Functions and Services

Ecosystem function	Ecosystem service
Regulation of atmospheric chemical composition	Gas regulation (e.g., CO_2/O_2 balance)
Regulation of global temperature, precipitation, and other biologically mediated climatic processes at global or local levels	Climate regulation (e.g., of greenhouse gases)
Capacitance, damping, and integrity or ecosystem response to environmental fluctuation	Disturbance regulation (e.g., storm protection, flood control, and drought recovery)
Regulation of hydrological flows	Water regulation (e.g., water for agricultural and urban uses)
Storage and retention of water	Water supply (i.e., provisioning of water by watersheds, reservoirs and aquifers)
Retention of soil within an ecosystem	Erosion control and sediment retention (e.g., prevention of loss of soil by wind, runoff, or other processes)
Storage, internal cycling, processing, and acquisition of nutrients	Nutrient cycling (e.g., nitrogen fixation)
Recovery of mobile nutrients and removal or breakdown of excess or xenic nutrients and compounds	Waste treatment (e.g., pollution control, detoxification)
Removal of floral gametes	Pollination
Trophic-dynamic regulations of populations	Biological control (e.g., keystone predator control of prey species)
Gross primary and secondary production	Food production (e.g., crops, hunting), raw materials (e.g., production of timber, fuel, and fodder)
Sources of unique biological materials and products, biodiversity	Biodiversity and genetic resources (e.g., medicines)
Habitat for algae, bacteria, fungi, fish, shellfish, wildlife, and plants	Recreation (e.g., vistas, forests), food production, biodiversity preservation
Providing opportunities for non-consumptive uses	Culture (e.g., educational, research, and spiritual values of ecosystems)

Source: adapted from Costanza et al. (1997) and Richardson (1994).

The application of ecological and social information, options, and constraints to achieve desired social benefits within a defined geographic area and over a specific period (Lackey 1998, pp. 23, 29).

Ecosystem management is best thought of as short-hand for "the process of ecosystem-based management of human activities." ... It is deliberate management of an entire regional ecosystem with the intention of maintaining ecological sustainability and/or integrity. Ecosystem-based management may often necessarily be a dispersed and collaborative activity, but the key is the focus on the whole ecosystem, defined in local, biophysical, and cultural terms, and on development of an integrative process for planning and management (Slocombe 1998, p. 483).

The Ecological Society of America ... identified eight common elements that are associated with ecosystem-based management:

1. Long term sustainability is a fundamental value;
2. Goals must be clearly defined;
3. Sound ecological models and understanding are essential;
4. Management efforts must recognize the complexity and interconnectedness of ecological systems;
5. Management efforts must recognize the dynamic character of ecosystems;
6. The design of management systems must be carefully crafted to suit specific local conditions;
7. Humans are a fundamental component of ecosystems; [and]
8. Knowledge of ecosystems is incomplete, ecosystems are dynamic, and a variety of changes occur over time, therefore, management should be adaptive and include a means of learning from policy experiments (Imperial 1999, pp. 451–452).

The ten dominant themes of ecosystem management are:

1. Hierarchical Context ... When working on a problem at any one level or scale, managers must seek the connections between all levels. This is often described as a "systems" perspective.
2. Ecological Boundaries. Management requires working across administrative/political boundaries ... and defining ecological boundaries at appropriate scales ...
3. Ecological Integrity ... Most authors discuss this as conservation of viable populations of native species, maintaining natural disturbance regimes, reintroduction of native, extirpated species, representation of ecosystems across natural ranges of variation, etc.
4. Data Collection. Ecosystem management requires more research and data collection ... as well as better management and use of existing data.

5. Monitoring. Managers must track the results of their actions so that success or failure may be evaluated quantitatively ...
6. Adaptive Management. Adaptive management assumes that scientific knowledge is provisional and focuses on management as a learning process or continuous experiment where incorporating the results of previous actions allows managers to remain flexible and adapt to uncertainty ...
7. Interagency Cooperation. Using ecological boundaries requires cooperation between federal, state and local management agencies as well as private parties. Managers must learn to work together and integrate conflicting legal mandates and management goals.
8. Organizational Change. Implementing ecosystem management requires changes in the structure of land management agencies and the way they operate. These may range from the simple (forming an interagency committee) to the complex (changing professional norms, altering power relationships).
9. Humans Embedded in Nature. People cannot be separated from nature ...
10. Values. Regardless of the role of scientific knowledge, human values play a dominant role in ecosystem management goals (Grumbine 1994, pp. 29–31).

Three features are striking in this handful of definitions and understandings: their considerable overlap; their abstraction and generality; and their question-begging. Just what is meant by ecosystem health, integrity, productivity, resilience, and sustainability? Just where is the management, let alone recoupling, of ecosystem functions and services?

It is a fairly simple matter to show that ecologists themselves differ over how to define the second-order terms and even over whether these terms can ever be defined satisfactorily (e.g., Lackey 1998). Substantial imprecision is to be expected, given the complexity and dynamics of ecosystems. These systems, as ecologists have pointed out (e.g., Grumbine 1997, p. 45; Gunderson 1999a), are moving targets with all manner of surprises. Delineating terminology here is about as easy as pinning down a butterfly in flight. Chapter 3 argues that ambiguity in terms is inevitable where notions of ecosystem integrity, health, sustainability, and resilience are rooted in presettlement or predisturbance conditions, which ecosystem decision makers are seeking to restore or mimic in their management efforts. Few, if any, adequate models exist for the presettlement template. Thus it is hardly unexpected that terminology for such conditions are found wanting (for a discussion of the different meanings of ecosystem health, integrity, biodiversity, and sustainability, see Slocombe 1998, pp. 486–487).

The lack of terminological clarity is one reason why ecologists insist that clear goals and objectives are key to effective ecosystem management.

Ecologists may disagree over whether a state of affairs constitutes fish recovery or habitat restoration, but if the specific objectives include ensuring survival of so many fish and so much habitat, then decision makers can at least evaluate whether ecosystem management has achieved its objectives. Focusing on specifics risks confusion between suboptimization and optimization. Nonetheless, in the highly charged task environment of species recovery and habitat restoration, clear means to ambiguous ends are often seen as a precondition for effective ecosystem management.

This is a necessary but insufficient precondition, though. Specific, measurable objectives are not ipso facto management-relevant, and the generality of the definitions is often linked to a palpable absence of specifics over how to undertake ecosystem management in the field. Indeed, one of the first things readers from public management or administration fields ask when immersing themselves in the ecosystem management literature is, "Where is the *management* in ecosystem management?" Entire books (e.g., Cortner and Moote 1999) have been written on ecosystem management without ever really discussing or specifying the nuts and bolts of the organizational management processes and field operations that decision makers must have in place in order to undertake ecosystem management. The reader has to search long and hard for "a day in the life of an ecosystem manager." Consequently, the drive to specific goals and objectives can be seen as much a response to the lack of specifics over implementation and operations as it is to the lack of terminological clarity in governing concepts.

We view the lack of management specifics in the literature as a blessing, however. The argument of our book is that it is far too early to talk about management specifics, as the framework, principles, and goals of ecosystem management require much more specification than is currently used. In our view, the reason why many of our case study interviewees feel that practice falls short of the theory in ecosystem management lies more with the theory than with the practice. We show that the overgenerality and underspecification of key points in the literature arise because relatively little attention has been given to the causal theories and analytical approaches for specifying options, tradeoffs, and priorities in ecosystem management. For the moment, though, ecosystem management is about options and their tradeoffs. If decision makers cannot be clear about those, then they are understandably left to wonder just what and how to manage.

Why, Then, Ecosystem Management?

If ecosystem management helps us to understand that functions and services should be managed together but is not specific enough to guide us in deciding how to do this, why should decision makers take ecosystem management seriously? Certainly, ecosystem management has ideological critics, on both the right and the left (e.g., Fitzsimmons 1999). Accusations of ecosystem management being just another buzz word, along with ecosystem health, sustainability, and the like, are common (see the review in Callicott et al. 1999). That noted, the literature offers at least four related reasons why ecosystem management is relevant for our book.

First, the status quo is itself taken to be a negative argument in favor of ecosystem management. Under current resource management policies and laws, ecosystem degradation continues on an unparalleled scale, with little or no sign that environmental deterioration is slowing down (Grumbine 1994, p. 29). A Policy Forum piece in *Science* sets out the boldest case yet for ecosystem management. In "International Ecosystem Assessment," Ayensu et al. (1999, pp. 685–686) offer the following dismal numbers:

> About 40% to 50% of land on the Earth has been irreversibly transformed (through change in land cover) or degraded by human actions ... More than 60% of the world's major fisheries will not be able to recover from overfishing without restorative actions ... Natural forests continue to disappear at a rate of some 14 million hectares each year. [Projections] suggest that an additional one-third of the global land cover will be transformed over the next 100 years. ... By 2020, world demand for rice, wheat, and maize is projected to increase by ~40% and livestock production by more than 60% ... Demand for wood is projected to double over the next 50 years.

Current approaches to resource management, which have been by and large sectorally based (for forests, or agriculture, or water), are inadequate to meet the challenges ahead. The older approaches "made sense when tradeoffs between goods and services were modest or unimportant. They are insufficient today, when ecosystem management must meet conflicting goals and take into account the interlinkages among environmental problems." What is required is an "integrated, predictive and adaptive approach to ecosystem management" (Ayensu et al. 1999, p. 685). Like so many authors before them, Ayensu et al. leave the reader wondering just how this approach would be managed and implemented on the ground.

The second reason for our taking ecosystem management seriously is that others do so as well. The concept of ecosystem management has been widely embraced by governments and nongovernmental organizations alike, most notably (but not exclusively) in the United States (Callicott et al. 1999). "At least in the political arena, the debate is concluded whether or not ecosystem management is a good idea; it *will* be implemented, or at least attempted, in word if not deed" (Lackey 1998, p. 22). According to Imperial (1999, p. 450), the U.S. federal agencies that have announced or engaged in ecosystem-based management activities include the Forest Service, the Bureau of Land Management, the Fish and Wildlife Service, the National Park Service, the National Resource Conservation Service, the Bureau of Reclamation, the Environmental Protection Agency, and the Department of Energy. Our research identified others as well. Terrestrial and aquatic ecosystems subject to ecosystem management include the Greater Yellowstone ecosystem, Columbia River Basin, San Francisco Bay–Delta, the Upper

Mississippi River, the greater Everglades, Chesapeake Bay, and Puget Sound (see Walters 1997; Imperial 1999; Johnson et al. 1999). Internationally, ecosystem management initiatives are increasingly common also (Gunderson et al. 1995).

Third, the science and conceptual foundations behind ecosystem management have substance (Lackey 1998, p. 22; Callicott et al. 1999). Those placing science-based, experimentally driven adaptive management at the core of ecosystem management are particularly instrumental in making this case (Holling 1978; Walters 1986; Walters and Holling 1990; Lee 1993). We have already seen the close connection between ecosystem and adaptive management. Others make the same link. For the U.S. Department of Agriculture, adaptive management is the "process of implementing policy decisions as scientifically driven management experiments that test predictions and assumptions in management plans, and using the resulting information to improve the plans" (United States Department of Agriculture, Forest Service et al. 1993, p. IX-1). Ecosystem management, in turn, is the "use of an ecological approach in land management to sustain diverse, healthy, and productive ecosystems ... [It] is applied at various scales to blend long-term societal and environmental values in a dynamic manner that may be adapted as more knowledge is gained through research and experience" (United States Department of Agriculture, Forest Service, and United States Department of the Interior, Bureau of Land Management 1994, Glossary-5). "Ecosystem management is adaptive because our knowledge is limited," according to Berry et al. (1998, p. 57). "Like many other government programs, ecosystem-based management is the result of an evolutionary process of experimentation, goal definition and redefinition, and the search for appropriate implementation strategies," concludes Imperial (1999, p. 460). "Management will be successful in the face of complexity and uncertainty only with holistic approaches, good science, and critical evaluation of each step," write Haney and Power (1996, p. 885), adding "Adaptive management is where it all comes together." Most commentators place adaptive management as centrally in ecosystem management (e.g., Grumbine 1994, p. 31).

Fourth, trying to preserve or save species and habitat piecemeal and ad hoc, especially through the species-by-species approach of the Endangered Species Act (ESA) and other legislation, has been much criticized and given rise to calls for a more multispecies, ecosystem-based, or whole-system approach to replace it (e.g., Grumbine 1994, p. 34; Wilson et al. 1994, p. 291). As one informed commentator put it for the most populated state in the United States, "California's resource managers are veterans of many endangered species battles. Several of these efforts have cost the parties involved years of negotiation, millions of dollars, and enormous amounts of political wrangling ... [W]ith more than 900 species in serious decline in California ... the prospect of species-by-species conservation was appalling" (Debra Jensen, ecologist with the California Department of Fish and Game, quoted in Thomas 1999). We return to the ESA literature later.

The central idea behind a whole-system approach is to preserve, restore or otherwise rehabilitate an ecosystem or landscape's functions and services so as to increase their resilience, the ability to bounce back from or absorb disturbances, particularly those that are exogenous in origin (on resilience, see Folke et al. 1996). A flood, for example, may destroy a spawning habitat of an endangered species, but management would aim to ensure that ecosystem functions are sufficiently resilient and self-sustaining to restore that habitat on their own.

What Are the Threats to Ecosystem Management?

The chief threats to ecosystem management are said to be institutional. These include but are not limited to legal restrictions such as those embodied in endangered or protected species legislation. The legally sanctioned, but often far wider, fragmentation of organizations and professions is treated by much of the literature as especially troublesome to ecosystem-based management (Slocombe 1998, p. 483). Surely the most repeated refrain in ecosystem management is some variant of "Responsibility [for natural resource management] is divided among a myriad of agencies ... [with] overlapping jurisdictions and differing mandates that often leave them at cross-purposes, which may impede efforts to adopt ecosystem management" (e.g., Cortner et al. 1998, p. 162). We hasten to add that such legal, regulatory, and jurisdictional constraints on ecosystem management are not peculiar to the United States. They are found in varying degrees of perniciousness in Europe and beyond (Röling and Wagemakers 1998).

The greatest threat to ecosystem management is, however, the failure to meet its greatest institutional challenge: the paradox. How are organizations to preserve, restore, and otherwise rehabilitate ecosystems, while ensuring the reliable provision of services (including goods) from those ecosystems? As Grumbine (1994, p. 31) states, "If ecosystem management is to take hold and flourish, the relationship between the new goal of protecting ecological integrity and the old standard of providing goods and services for humans must be reconciled." Our book demonstrates how meeting this promise is possible but in no way guaranteed. There is reason for hope, though. It is said that some 90% of all scientists that ever lived are alive today, such that the pace of innovation does not just seem faster, it really is faster. So too we believe for the innovations in meeting the twofold goal. Never in history have there been as many ecologists and engineers alive as today. Never more than now can we expect them to produce innovations in the ways they deal with their respective systems collaboratively.

What Does Ecosystem Management Promise Specifically?

Ecosystem management preserves, restores, and rehabilitates ecosystems. But what is an "ecosystem" and just what do "preservation," "restoration," and "rehabilitation" mean? Conventional definitions of the first are much in the style of Odum's: "Any unit that includes all of the organisms in any given area interacting with the physical environment so that a flow of energy leads to a clearly defined trophic structure, biotic diversity, and material cycles is an ecological system or ecosystem" (Odum 1971 in Berry et al. 1998, 56). The definition is wide enough to include humans. But only recently has the notion of human-dominated ecosystems come to the fore (see Matson et al.; also Vitousek et al. in the *Science* 1997 special section on human-dominated ecosystems; see also Strange et al. 1999). In these ecosystems, humans are not just one among many different organisms interacting with a wider environment. Humans dominate the ecosystems and that environment, either by commission or omission. People have changed and transformed ecosystems in irrevocable and unintended ways. Thus it is they who must preserve, restore, or rehabilitate what is left of the ecosystems.

Because these are human-dominated ecosystems, political, social, and cultural values are as intrinsic to them as are the processes of photosynthesis, denitrification, and organic-matter accumulation. These values—all of which have a historical dimension (Scoones 1999)—are an integral part of any ecosystem management (rather than repeating the literature, see Grumbine 1994; Berry et al. 1998; Lackey 1998; Cortner and Moote 1999). They set the boundaries for management. In fact, they enable the existence of ecosystem management and define which options, tradeoffs, and priorities are deemed feasible. Adaptive management, for example, can look very different depending on where you are in the world. Ashby et al. (1995) describe adaptive management-like interventions in Colombia where "scientists" are the farmers themselves, in contrast to U.S. initiatives, which involve many more professionally trained scientists (e.g., McLain and Lee 1996; see also Matson et al. 1997). Even the high reliability management of natural resources reflects and promotes different values depending on where you are. Pastoralism exhibits precisely the same core features of high reliability management (Roe et al. 1998) as high reliability organizations in Western countries. Yet the latter value technocratic control of the relevant task environment in ways that pastoralists do not.

For our purposes, the core values of specific import are the wider commitments to preserve, restore, or rehabilitate ecosystems for current and future generations. Restoration, in particular, is an increasingly topical commitment (*Science* 1999). Although commonly used and understood, the terms preservation and restoration are misleading: what is being preserved is thought to be something that is already wild or natural; and what is being restored is to be brought back to its natural or wild state. As such, restoration has been defined as "the complete structural and functional return of a biophysical system to a predisturbance state" (Rhoads et al. 1999, p. 304). Yet ecologists argue that "no ecosystem on Earth's surface is free of pervasive human influence" and there "is no clearer illustration of the extent of human dominance of Earth than the fact that maintaining the diversity of 'wild' species and the functioning of 'wild' ecosystems will require increasing human involvement" (Vitousek et al. 1997, pp. 494, 499; see also Strange et al. 1999). From this vantage point, nothing on the Earth's surface is really wild or natural' save some high ice caps and deep seabeds. So nothing can be preserved or even restored back to such a state. Indeed, one major consequence of living in human-dominated ecosystems has been the problematizing of all things "natural." For algae, eutrophication is a great thing; for many humans, it is not. Ants, termites, and insects continue to influence the vast majority of the planet's land surface, no matter how humanly altered or socially constructed it is. Is that unnatural?[1] "How can we be against agricultural biotechnology," a crop breeder asked one of us, "when the gene is the most natural thing around?"

Similar definitional problems arise with "rehabilitation" and related concepts. According to Rhoads et al. (1999, p. 304), the National Research Council defines "rehabilitation [as] a partial structural or functional return to the predisturbance state, and enhancement [as] any functional or structural improvement, a definition that is inherently tautological and therefore not useful in any practical sense." A more charitable view of the terminological confusion is that "restoration" and "rehabilitation" (and associated terms like enhancement, mitigation, refoliation, and reclamation), while often used inconsistently, nonetheless have real meaning in practice. In contrast to the definitions in the preceding paragraph, restoration more often means reverting to the extent possible to historic conditions, while the more realistic option may be to rehabilitate the ecosystem by reintroducing something *like* the complexity and unpredictability it once had or by improving, say, specific habitat for a high-valued species that it might never have had historically (see Callicott et al. 1999 for a detailed discussion of distinctions between restoration and rehabilitation). In the latter case, rehabilitation could enhance activities that encourage exotic species, while restoration would encourage the return of native species only. These need not be mutually exclusive, however. Callicott et al. (1999, p. 28) give an example where "the ecological rehabilitation by use of exotic species accidentally contributed to the restoration of some elements of Lake Michigan's 'original' biotic community." Still, the two terms are often collapsed together (e.g., the 1999 CALFED Strategic Plan for Ecosystem Restoration quoted in chapter 6).

Alternative terminology to restoration and rehabilitation, such as "naturalization," has its own problems. Rhoads et al. (1999, p. 304) argue that, in contrast to restoration or rehabilitation, the "goal of [stream] naturalization is to establish sustainable, morphologically and hydraulically varied, yet dynamically stable fluvial systems that are capable of supporting healthy, biologically diverse aquatic ecosystems." Unclear is how this differs from what is actually done in many initiatives under the name of restoration and rehabilitation. Accordingly, this book continues to use the terms, "preservation," "restoration," and "rehabilitation," with the qualification, however, that the predisturbance state is no longer the primary guide to any of those enterprises. As Light et al. (1995, p. 104) said of efforts to restore the Everglades ecosystem, "By *restoration*, we do not mean returning the system to the way it used to be, but rather renewing its vitality by reuniting the systems functions." The question of course is, how do you manage to recouple these functions and their services?

What Needs to Be in Place for Ecosystem Management?

From the perspective of many commentators, fulfilling ecosystem management's promise requires organizational change, including but not limited to more effective interagency coordination to deal with problems of fragmentation. While our research confirms that interagency collaborations have produced appreciable results, there are no recipes here. Nor could there be for something as complex as ecosystems. The literature is clear, however, that more than organizational change is needed if ecosystems are to be better managed. Also called for are better pricing, broad institutional reforms, and targeted incentives. Indeed, many environmental economists and others, including ecologists, see these interventions as especially necessary, albeit some people doubt that the invisible hand has a green thumb (e.g., Ayres 1995, p. 98; Page 1995, p. 141; Slocombe 1998, table 3). Advocates, for example, argue that

> while economic *growth* may not ameliorate environmental problems, economic *development* often can. Economic development broadly defined includes the reduction of price distortions (e.g., underpricing of energy), efficiency-promoting macro-level adjustments (e.g., reductions in credit scarcity that limit natural resource destruction, and modification of import substitution policies that protect existing harmful activity while discouraging innovative new investment), and institutional reforms (e.g., safeguards for private and collective property that are likely to enhance protection of ecological resources. (Toman 1996, p. 136)

> The solution to environmental degradation lies in such institutional reforms as would compel private users of environmental resources to take account of the social impacts of their actions. ... These institutions need to be designed so that they provide the right incentives for protecting the resilience of the ecological systems. (Arrow et al. 1995, pp. 92, 94)[2]

Specifics on how to undertake such interventions are just as scarce as ways to ensure better interagency coordination. The broad appeal to prices, institutions, and incentives, without elaborating the details of actual implementation and operations, reinforces the overgenerality and underspecification of the ecosystem management literature noted earlier. It is as if management will take care of itself once the overarching market and institutional structures are in place, even though the now-vast organizational literature speaks with one voice: actual implementation consistently falls short of stated policy goals, thus rendering implementation and operational management as a kind of de facto policymaking. Ludwig (1996), Schindler (1996) and other ecologists are surely correct when concluding that, if there is anything certain in ecosystem management, it is that good management is never automatic in real time.

Not only must better coordination, prices, institutions, and incentives be in place, but better professionals are needed as well. With disciplines, fields, and professions come self-selection and trained incapacities (what we call "well-earned blind spots" in chapter 6). Professionals, no matter how competent, seek to work with others who share their worldviews and thus view only part of the same world. "Many ecologists have a strong tendency to support 'environmentalist' worldviews and positions," Lackey (1998, p. 28) notes, adding that this "is understandable, in part due to self selection in all disciplines (environmentally oriented individuals are more likely to select ecologically oriented fields than are more materially oriented individuals) . . . Individuals, in any profession, naturally tend to be advocates for what is important in that profession." Yet with increasing complexity also comes increasing loss of competence. Cognitive incomprehension rises, and experts find it more and more difficult to practice their expertise precisely because matters have become increasingly uncertain and complex. According to Sartori (1989, p. 393), "My sense is that we are in fact backsliding, for we do not know *how to do* the very things we are doing. What we have is engineering without engineers." What we have is ecosystem management without ecosystem managers, at least in the view of some commentators.

Finally, Who Are the Professionals and Stakeholders in Ecosystem Management?

Just who are these professionals in ecosystem management? Are they the only ones that matter? Our book's answer to this question is different from the one commonly given. A prevailing view is summarized by Gretchen Daily, Paul Ehrlich, and Marina Alberti (1996, pp. 19–21) in their "Managing Earth's Life Support Systems: The Game, The Players, and Getting Everyone to Play." Since the issue is an important one to ecosystem management and central to our book, we spend some time on the Daily et al. position and its implications.

The authors focus on which professionals are to do what in ecosystem management, which they call sustainable Earth management. First, there are the ecologists, broadly understood. According to the authors, "the message is clear: the collective expertise of ecologists, climatologists, toxicologists, oceanographers, hydrologists, indeed most natural scientists is needed to characterize the human impact that can be sustained by [the Earth's] natural systems. In essence, a primary role for natural scientists is to *elucidate nature's rules of* [the] *game* and communicate them to others" (Daily et al. 1996, p. 20).

After ecologists come the economists:

> Economists are needed to help society determine what constraints on growth are necessary to provide reasonable insurance against eco-catastrophe, and what the most economically efficient and socially acceptable way is to obtain the necessary coverage. ... In short, economists are needed *to find the best strategies for winning the game*—how to best convert Earth management goals into policy to achieve them. (Daily et al. 1996, p. 20)

Then come the technicians:

> Technologists, i.e., agronomists, engineers, architects, city planners, and many others, are needed to create new, innovative tools and plans with which to produce/capture/convert/arrange the material ingredients of human well-being. They have the expertise to *develop superior tactics for winning the game*—how to most efficiently supply food, fiber, water, energy, communication, transportation, etc., to maximize the human benefit derived from each unit of environmental impact. (Daily et al. 1996, p. 20)

This ambitious enterprise needs other professionals as well. "Finally, the help of other disciplines, such as political science, other behavioral sciences, and the law, will be necessary to *get people and nations to play the game.*" Fortunately, so say the authors, many of the experts are already in universities and available for the task ahead. "We hope cooperation will lead to further transdisciplinary collaboration at the interface of scholarship and environmental policy" (Daily et al. 1996, p. 20).

Many ecologists believe, perhaps more privately than publicly, that some such hierarchy of roles is in order, and we thank Daily et al. for putting this belief into print. That said, the hubris of the position is breathtaking. The experts are in charge and they are not us; the hierarchy is a linear and tightly coupled chain of control. Success is in sight if only we would let ourselves be educated by those who know best. Science is the polestar, ecologists are in power (finally!), economists are their chief strategists, engineers and operators are the get-dirty tacticians, and the policy types are there to get and keep the masses in line. As the abundant literature referred to a moment ago on governance, organizations, and implementation reiterates, such "planning" has not worked, will not work, and cannot work. The decision space in which ecosystem decision makers operate is far too complex for blueprint management to hold, a point detailed in chapter 4. For the moment, note only that the implementation and operations literature underscores that implementers, including the engineers and line operators who play an important part in the rest of this book, are key in effective ecosystem management and the policymaking around it. To their great credit, we believe Daily et al. have the players right, but the order dangerously wrong.

Nor must it be only professionals who are active in ecosystem management. Stakeholders are more and varied (Smith et al. 1998). In the view of one ecologist interviewed for this book,

> in practice, ecosystem management and adaptive management have meant that the power of [the planning and regulatory] agencies has been reduced. That is, for years and years the ethic was, when in doubt, turn to the technocrats to solve your problems. I'm wondering if that first wave of ecosystem management and adaptive management is about trying to share that power more with stakeholders. I think that has actually happened. Power has become more distributed.

Often these stakeholders are themselves professional or expert groups, as in non-governmental organizations representing specific species interests or the interests of various water communities, though such organizations can be citizen groups as well (e.g., Grumbine 1994, p. 32).

While short on specifics, much of the literature is in agreement that stakeholders matter, and they matter early on and throughout the ecosystem management process. Referring to an interagency ecosystem management project in the Columbia River Basin, one theme of those interviewed in Berry et al. (1998, p. 67) was that "[o]utstanding public involvement at all stages improved every aspect of the work of the Project. Involvement and communication have resulted in notably increased public trust and confidence" (see also Grumbine 1997; Lackey 1998; Imperial 1999). Nor is the call for more active involvement of stakeholders in ecosystem management surprising, given the plural nature of the political, social, and cultural values that inform its goals, objectives, and priorities. Discussions about different values on the one hand and different stakeholders on the other are often proxies for each other (e.g., Cortner et al. 1998).

Like much of the literature (e.g., Grumbine 1994; Thomas 1997), we argue that ecologists must be taken more seriously as stakeholders than they currently are, but in ways that require changes on their part and on the part of their key collaborators, particularly engineers and line operators working in the large-scale water and power systems that currently dominate ecosystem services. Paul Risser (1996, p. 25), an ecologist with strong administrative bona fides, has argued that "as ecologists, we will have a significant impact only when we move past or apply the general theories to the specific arenas where decisions are made. Our success will require some fundamental changes in the institutional organization of our science and in the ways that we establish the criteria for evaluating our research contributions and society."

While the ecosystem management literature is much less explicit, our interviews suggest that engineers and line operators are also key stakeholders—if only because many high reliability organizations provide not only water and power, but are increasingly required to undertake habitat restoration. As our interviews make clear, wherever ecosystem management is undertaken in a context that values service reliability, there too engineers and line operators play an active role. While they are part of the problem, they are also a major part of resolving the paradox, along with ecologists. Even ecologists recognize this. In the words of one interviewee,

> it will take a lot of creativity to balance the technological and engineering solutions with the processes of natural systems. In the past we often engineered a substitute for the degraded system, such as the hatcheries. We have responded to the ecological problems with technology. Put fish in barges and take them down the river. It doesn't work. There needs to be a blending: technological solutions that take into account the natural processes, such as innovative hatcheries. At the same time you can't send people back to Europe.

In a nutshell, the quote states the need for a better blending of engineering and ecology, even though the cultural differences mean engineers and ecologists overcoming their blind spots. The former has, as one of our interviewees put it, "an orientation towards quantifying and analyzing" the problem, while the latter are more oriented to "qualitative notions of system health." Along the same lines, Holling (1996, p. 33), a well-known ecologist, characterized the different views of resilience: the engineering "definition focuses on efficiency, constancy, and predictability—all attributes at the core of the engineers' desires for fail-safe design. The [ecological definition] focuses on persistence, change, and unpredictability—all attributes embraced and celebrated by biologists with an evolutionary perspective and by those who search for safe-fail designs." Chapter 6 discusses these blind spots in detail. Equally important, the way in which the engineering literature frames the link between ecosystem services and functions is discussed under the topic of high reliability management of natural resources in chapter 4.

What is the role of economists as stakeholders? In the literature, theirs are among the most dominant views. "Economists have a much greater role in, and influence on, policy debates than do ecologists," concludes Michael Common in his "Economists Don't Read Science" (Common 1995, p. 102). We, however, do not see the dominance reflected in the literature on concrete ecosystem management initiatives nor in the interviews undertaken for this book. The interviews directed our attention much more on the impasse between ecologists and engineers (broadly writ). We followed this empirical lead by focusing on that impasse at the core of the paradox. No doubt economists will play an increasingly important role in ecosystem management. By matching the appropriate management regimes to their respective ecosystems, it will become much clearer how their generic proposals for better prices, institutions, and incentives can be applied case by case.[3]

Finally and more generally, ecosystem management poses a fundamentally different way of seeing stakeholder involvement than is currently the case in much of natural resource management. Instead of management being top-down by experts versus bottom-up from the public and their values (e.g., see Thomas 1997), stakeholder involvement can be seen as outside-in versus inside-out management (Roe 1998). Assume for illustrative purposes that the planning requirement for ecosystem management is that all the stakeholders in the ecosystem are at the table to hammer out a management plan for the area in question (indeed a requirement of some legislation). Assume the stakeholders are developers, environmentalists, local leaders, and expert officials in the state regulatory agency. What if the developer owns land in the area but actually lives and works elsewhere; the environmentalist visits the area only on weekends; and officials travel from the capital to the area only for meetings? They may be *de jure* stakeholders, but do they have the same *de facto* "stake" in the ecosystem as local community leaders and residents? Are all stakes equal? Just what is the "local community," when the distinction between insiders and outsiders is becoming less and less clear? Increasingly, the challenge in ecosystem management is to come up with varieties of inside-out planning for ecosystem management, where local leaders and residents are themselves the experts and where the planning process is itself initiated and guided from within the local ecosystem. As we will see in later chapters, the development of user-friendly ecological and environmental models has enabled those outside the professions to become "amateur" ecologists or engineers. Ecosystem management offers other prospects for increasing local people's expertise, though not without difficulty (see Grumbine 1994).

We turn now to a brief introduction and review of the three case studies that form the heart of our book: ecosystem management initiatives in California's San Francisco Bay–Delta, the Pacific Northwest's Columbia River Basin, and Florida's Everglades. The final chapter of the book introduces a fourth case study, the Netherlands' Green Heart area, which is used to study the paradox in a different context. For now, our attention is directed on the three U.S. initiatives. By the end of the book and its case studies, we will have developed our own answers to the questions touched on in this chapter.

Introduction to Case Studies

The three U.S. ecosystem management initiatives differ in many respects, but also share important similarities that make them suitable cases for an empirically guided understanding of the paradox. The most important similarity is the conjunction of three features. Each initiative: (1) focuses considerable attention and resources on improving and recoupling ecosystem functions and services, though variously termed; (2) relies on adaptive management, though in different ways, to do so; and (3) involves state and federal agencies that are responsible not only for ecosystem rehabilitation but also the provision of reliable services, such as water supply, water quality, and power generation. In one sentence, the three cases are in zones of conflict. These points are introduced in more detail in chapter 3. The features make the case studies key in deriving answers to the book's main questions.

All three ecosystem management initiatives are directed at major aquatic ecosystems. We focus on aquatic ecosystems because they reflect the most difficult and enduring management issues. Many would agree that our knowledge about them is more uncertain and incomplete than that on terrestrial ecosystems. Here, in other words, the paradox appears most intractable and thus most important for exploring the opportunities for its resolution. Aquatic ecosystems also feature strongly in the adaptive management literature. The initiatives for the Everglades and the Columbia River Basin are arguably the two most cited cases, while many case studies in that literature are on aquatic ecosystems, both inside and outside the United States (e.g., Lee 1993; Light et al. 1995; McLain and Lee 1996; Walters 1997; Gunderson 1999a; Johnson et al. 1999).

While the cases enable us to learn about the paradox, it is also vital to understand the differences between them. Table 2.2 compares some main characteristics reported by the initiatives. As one might expect, data isolated from their context are difficult to compare. The table, however, gives a sense of salient differences between the cases.

There are other ecosystem management efforts occurring in these geographical areas which will not be discussed. However, given their importance to the greater Columbia River Basin, the activities of the Bonneville Power Administration and the Northwest Power Planning Council related to ecosystem restoration and service reliability are discussed in detail, here and in subsequent chapters.

The cast of major characters involved in the initiatives must be introduced before describing each case. Many federal and state agencies have played an important role in the three initiatives. Table 2.3 introduces the main participating federal agencies, which reappear in subsequent chapters. The state cast of characters varies considerably across the three case studies; state counterparts to the federal agencies are not everywhere present or equally important. The relevant state agencies and groups are identified in the following sections and discussed more fully in subsequent chapters.

CALFED Bay–Delta Program

The CALFED Bay–Delta Program has been a combined effort among California state and U.S. federal agencies to address environmental and water management problems associated with the San Francisco Bay/Sacramento–San Joaquin Delta estuary (Bay–Delta). The San Francisco Bay–Delta is a web of waterways created at the junction of the San Francisco Bay and the Sacramento and San Joaquin rivers and the watershed that feeds them (figure 2.1).

The Bay–Delta's maze of tributaries, sloughs, and islands supports more than 750 plant and animal species and covers an estimated 11,500 square

Table 2.2 Selected Characteristics of Three U.S. Case Studies

	San Francisco Bay–Delta*	Columbia River Basin**	Everglades***
Initiative	CALFED Bay–Delta Program	Interior Columbia Basin Ecosystem Management Project	Comprehensive Everglades Restoration Plan
Geographic area of initiative (appr.)	61,000 square miles (watershed); 11,500 square miles (Delta)	225,000 square miles	18,000 square miles
Estimated human population in the area	Over 7 million in the 12-county Bay–Delta area	Over 3 million	Over 6 million
Key ecosystem services in the area	Drinking water to 22 million people and irrigation water to 7 million acres of the most productive farmland in the world	Timber production, agriculture, and thousands of dams for water storage, irrigation, flood control, and production of 40% of total U.S. hydropower	Drinking water, irrigation, and flood protection to 6.3 million people, major ecosystem-related recreational services
Number of plants and animal species identified in the area	Over 750 (Delta)	Over 17,000	"thousands"
Number of ESA-listed species	88 (58 endangered, 30 threatened)	38 (19 endangered, 19 threatened)	68 (51 endangered, 17 threatened)
Estimates cost of implementation of preferred alternative	$9 billion to $10.5 billion	$202 million per year	$7.8 billion (plus $182 million annually for maintenance and operation)
Estimated implementation time	30 years or more	10–15 years	Over 20 years
Number of responsible	18 lead agencies	2 lead agencies, several partners	2 lead agencies, over 30 partners

* Data derived from CALFED (1999a) and the CALFED website (http://calfed.ca.gov/general/about_bay_delta.html and http://calfed.ca.gov/general/new_q&a.html; November 17, 2000).
** Data derived from United States Department of Agriculture, Forest Service, and the United States Department of the Interior, Bureau of Land Management (1997, 2000c). The data on ESA listings was kindly provided by Barbara Wales, Forest Service, PNW Forestry and Range Sciences Lab, La Grande, Oregon (2000, pers. comm.).
*** Data derived from fact sheets on the Comprehensive Review Study website (http://www.evergladesplan.org/pub_Restudy_EIS.htm; November 17, 2000). Data on ESA listings is derived from United States Department of the Interior, Fish and Wildlife Service (1999, pp. xii–xvi).

Table 2.3 Main Federal Agencies in U.S. Case Studies

Name	Brief description
Bureau of Reclamation	BR (Department of the Interior) is a water management agency, best known for the dams, power plants, and canals it constructed in 17 western states. It has constructed more than 600 dams and reservoirs and is the second largest wholesaler of drinking and irrigation water in the country. BR is the second largest producer of hydroelectric power in the western United States.
Bureau of Land Management	BLM (Department of the Interior) manages over 260 million acres of surface acres of public lands located primarily in 12 western states. The agency manages an additional 300 million acres of below-ground mineral resources. Its original focus was commodity and service production. Its management covers energy and minerals, timber, forage, fish and wildlife habitat, wilderness areas, and sites with natural heritage values.
Fish and Wildlife Service	FWS (Department of the Interior) is the principal federal agency responsible for conserving, protecting and enhancing fish, wildlife, plants, and their habitats. It enforces federal wildlife laws, protects endangered species, manages migratory birds, restores nationally significant fisheries, conserves and restores wildlife habitat, such as wetlands.
National Park Service	NPS (Department of the Interior) seeks to preserve the natural, recreational, and cultural resources of the 380 areas it manages. The areas cover more than 80 million acres in all but one state and include national parks, monuments, battlefields, military parks, historical parks, historic sites, lakeshores, seashores, recreation areas, and scenic rivers and trails.
U.S. Geological Survey	USGS (Department of the Interior) provides geologic, topographic, and hydrologic information for the management of natural resources. The information consists of maps, data bases, and descriptions and analyses of the water, energy, and mineral resources, land surface, underlying geologic structure, and dynamic processes of the Earth.
Environmental Protection Agency	EPA is a regulatory agency that coordinates governmental action to protect and manage the environment. EPA is responsible for research, monitoring, standard setting, and enforcement activities for the legally mandated goals of ensuring clean air and water, safe food, prevention and reduction of pollution, risk management, and waste management.

Table 2.3 (*continued*)

National Marine Fisheries Service	NMFS (Department of Commerce) is a regulatory agency in the National Oceanic and Atmospheric Administration that is responsible for protecting marine species, among which are those listed under the Endangered Species Act. NMFS supports the development of a commercially viable and environmentally sound domestic aquaculture industry.
Army Corps of Engineers	ACE (Department of Defense) provides engineering services, most notably for the planning, designing, building and operating of water resources and other civil works projects, including dams and levees. It has a diverse workforce of biologists, engineers, geologists, hydrologists, natural resource managers and other professionals.
Bonneville Power Administration and Western Area Power Administration	BPA (Department of Energy) wholesales electric power produced at 29 federal dams located in the Columbia–Snake River Basin. Its service territory includes Oregon, Washington, Idaho, and Montana and it sells surplus power to California and the Southwestern United States. BPA is a federal utility, one of five power marketing agencies. The others are the Western Area Power Administration and the Southeast, Southwest, and Alaska power administrations. Legislation extended BPA's responsibilities to include improvement of fish and wildlife resources that have been affected by the construction of hydropower plants.
Forest Service	FS (Department of Agriculture) manages public lands in national forests and grasslands. It is responsible for protection, management, and production of natural resources on national forest system lands. It cooperates with state and local governments, forest industries, and private landowners to help protect and manage non-federal forest and associated range and watershed lands.

Sources: All descriptions have been adapted from online sources. Original material can be found for: BR (http://www.usbr.gov/main/what/who.html); BLM (http://www.blm.gov/nhp/faqs/faqs1.htm#1); FWS (http://www.fws.gov/r9extaff/pafaq/fwsfaq.html); NPS (http://www.nps.gov/pub_aff/e-mail/faqs.htm); USGS (http://www.usgs.gov/usgs-manual/120/120-1.html); EPA (http://www.epa.gov/ocfopage/plan/2000strategicplan.pdf); NMFS (http://www.nmfs.noaa.gov/om2/contents.html); ACE (http://www.usace.army.mil/who.html#Mission); BPA and WAPA (http://www.bpa.gov/Corporate/KC/palinksx/faqsx.shtml); FS (http://www.fs.fed.us/intro/meetfs.shtml).

miles in five counties. While the program is first and foremost directed at the problems in the Delta itself, it has designated the wider watershed, draining more than 37% of the state (61,000 square miles), as its solution area. According to CALFED documentation, the scope of possible solutions to the major ecosystem related problems encompasses "any action that can be implemented by the CALFED agencies, or can be influenced by them, to address the identified problems—regardless of whether implementation takes place in the Bay–Delta" (CALFED 1999a, p. 1-10).

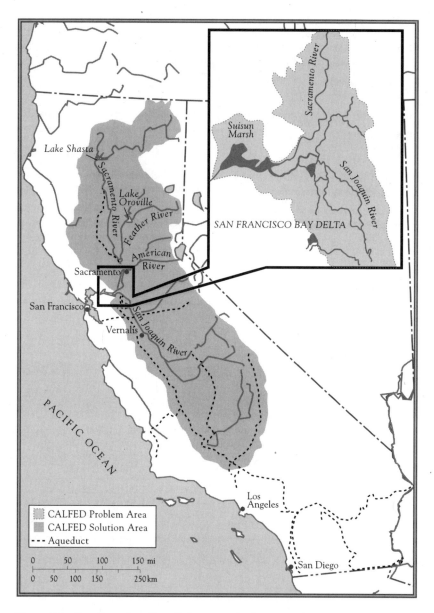

Figure 2.1   San Francisco Bay–Delta watershed.

The Bay–Delta has been dramatically altered by humans in the last two centuries. From an original 540 square miles, only 13 square miles of Delta tidal marsh remain. Waterways and islands are protected by more than 1,100 miles of levees. Some Delta islands have subsided to such a degree that they are now 20 feet below sea level. It has been estimated that two-thirds of the state's rain falls in northern California, while two-thirds of the people reside southern California, an asymmetry used to justify the massive, reliability-driven water conveyance infrastructure of the California State Water Project and the federal government's Central Valley Project. The Bay–Delta is the hub of these two water distribution systems, which together divert on average around 25% of the total inflow of water into the Delta—though this percentage fluctuates erratically over time.[4] In addition to the two systems, over 7,000 permitted diverters receive water supplies from the watershed feeding the Bay–Delta estuary. The diversions, along with the introduction of nonnative species, water contamination, and other factors have caused serious harm to the fish and wildlife of the estuary.

For decades, the Bay–Delta has been the focus of competing economic, ecological, urban, and agricultural interests. A variety of environmentalist nongovernmental organizations seek to conserve the largest estuary on the Pacific side of North and South America, home to a reported 130 fish species and millions of local and migratory birds. Anglers and commercial fishers are concerned about the sustained use of one of the most productive natural salmon fisheries on the American West Coast. California's agricultural industry, which at the end of the last century accounted for nearly $25 billion per year, demands the supply of irrigation water to millions of acres of the world's most productive farmland (in-Delta production by itself accounted for $500 million). Currently, some 12 million users enjoy camping, boating, fishing, and other recreational amenities each year. More than 22 million Californians rely on the Bay–Delta system for all or some of their drinking water. An estimated 4,000 commercial ocean-going vessels account for a total of more than 50 million tons of cargo a year in the San Francisco estuary, providing some 80,000 jobs and over $10 billion in annual revenues to Bay Area counties. As found in the other case studies, there are variety of stakeholder groups representing each of these interests.

Against this background, the CALFED Bay–Delta Program started in 1995 as a mechanism to address the complex management issues surrounding the use of the Bay–Delta resources, end decades of conflict among stakeholder groups and vested interests, and foster more cooperation among the 18 California state and federal ("CALFED") agencies having management or regulatory responsibilities in the region. The Program's institutional structure consists of an Executive Director, with appointed staff from the 18 participating agencies (table 2.4). The Program has consulted with the Bay–Delta Advisory Council, the citizen's advisory committee representing different stakeholder groups, and with representatives from the Native American tribes in the program area.

Among the state agencies involved in CALFED, the California Department of Water Resources (DWR) is a key player. It provides water for municipal, industrial, agricultural, and recreational uses; manages the control room of the State Water Project (next to the BR's control room of the Central Valley Project); regulates dams and reservoirs; and provides flood protection and emergency management. It is mandated to protect and restore the Bay–Delta and wider watershed by controlling salinity and providing water supplies for

Table 2.4 The CALFED Participating Agencies

State	Resources Agency of California
	Department of Water Resources
	Department of Fish and Game
	Reclamation Board
	California Environmental Protection Agency
	State Water Resources Control Board
	California Department of Food and Agriculture
	Delta Protection Commission
Federal	Department of the Interior
	Bureau of Reclamation
	Fish and Wildlife Service
	Bureau of Land Management
	U.S. Geological Survey
	Environmental Protection Agency
	Department of Commerce
	National Marine Fisheries Service
	U.S. Army Corps of Engineers
	Western Area Power Administration
	Department of Agriculture
	Forest Service
	Natural Resources Conservation Service

Source: CALFED (1999c, p. 5).

water users, by planning long-term solutions for environmental and water use problems facing the Delta, and by administering levee maintenance and special flood control projects. Important for understanding the case material, DWR works with local water agencies such as the Contra Costa Water District. Its mission, like that of other water districts, is "to strategically provide a reliable supply of high quality water at the lowest cost possible, in an environmentally responsible manner" (Contra Costa Water District 2000). The most famous water district, and as such also a key player in the consortium, is the Metropolitan Water District of Southern California (MWD), whose service region includes the greater Los Angeles area.

The need for a consortium of agencies reflects that, while DWR is a major player, ecosystem restoration and water reliability are under the purview of other state and federal agencies also (table 2.3). The California Department of Fish and Game (DFG) manages state lands, including wildlife areas, ecological reserves, and public access sites. In addition, DFG reviews environmental documents for land and water projects that may affect fish and wildlife. The State Water Resources Control Board (SWRCB) is a regulator seeking to ensure high water quality in the state. It has joint authority for water allocation and water quality protection and thus is concerned both with aquatic ecosystem health and service reliability.

Though improved coordination of agencies is an important aspect of CALFED, its raison d'être arises out of extreme resource conflict. Consequently, CALFED "solution principles" include statements that "solutions will reduce major conflicts" and "improvements for some problems will not be made without corresponding improvements for other problem." (CALFED 1999a, p. ES-5). CALFED identified several key threats to the Bay–Delta resources: declining fish and wildlife habitat; native plant and animal species threatened with extinction; degradation of the Delta as a reliable source of high quality water; and a Delta levee system faced with a high risk of failure. Consequently, CALFED arrived at four "Primary Objectives" to be pursued at the same time in order to create a "win-win" resource policy. The objectives have been variously summarized as "ecosystem quality," "water supply reliability," "water quality," and "levee system integrity." The most recent details on CALFED objectives are quoted in table 2.5.

Program execution is to be in three phases. Phase I, completed in August 1996, concentrated on identifying and defining the problems confronting the Bay–Delta system. The mission statement and guiding principles were developed, along with program objectives and an array of potential actions to meet them. Relying on a number of different models, both engineering and ecological, CALFED developed its "Preferred Program Alternative" in phase II, together with an implementation plan for the first seven years of the next phase. Phase II was concluded in 2000 with the completion and certification of a comprehensive programmatic environmental review of the preferred alternative—more formally, the Environmental Impact Statement/Environmental Impact Report (EIS/EIR). Site-specific, detailed environmental reviews will occur during phase III, prior to the implementation of each proposed action. Implementation is expected to take 30 years or more.

During Phase I, it was decided that the four program elements were so fundamental to the system's recovery that they should be included in whatever solution was ultimately chosen. The "common programs" are ecosystem restoration, levee system integrity, water quality and water use efficiency. During phase II, four additional common program elements were added based on public input and technical analysis: watershed management, water transfers, storage, and conveyance. These program elements, however, do not supersede the importance of the first four.

Table 2.5 Four Primary Objectives of the CALFED Program

- Ecosystem quality: improve and increase aquatic and terrestrial habitat and improve ecological functions in the Bay–Delta to support sustainable populations of diverse and valuable plant and animal species.
- Water supply: reduce the mismatch between Bay–Delta water supplies and the current and projected beneficial uses dependent on the Bay–Delta system.
- Water quality: provide good water quality for all beneficial uses.
- Vulnerability of Delta functions: reduce the risk to land use and associated economic activities, water supply, infrastructure, and the ecosystem from catastrophic breaching of Delta levees.

Source: CALFED (1999a, pp. 1–5).

The first common program is the Ecosystem Restoration Program (ERP), which is intended to provide significant improvements in habitat for the environment, to restore critical water flows, and to reduce conflict with other Delta system resources. At the time of writing, ERP contained over 700 programmatic elements to be implemented over the 30-year life of CALFED. The institutional framework under which the ERP will be carried out has been uncertain and is discussed in chapters 5 and 6. The "levee system integrity" component, or Long-Term Levee Protection Program (hereafter, "Levee Protection Program"), refers to the major structural improvements needed to enhance the reliability of Delta levees and avoid catastrophic failure from events such as flooding, subsidence, and earthquakes.

The Water Quality Program has been aimed at making significant reductions in point and nonpoint source pollution, thereby increasing the reliability of water quality for urban, environmental and agricultural uses, among other legally mandated beneficial uses. Last but not least, CALFED's Water Use Efficiency Program is meant to provide policies for efficient use and conservation of water in agricultural and urban settings and for environmental purposes, where such actions could alter the pattern of water diversions.

Reliability issues and mandates are clearly present throughout the core elements of the Program, be it restoration, levee improvement, water quality or water supply. CALFED has also elevated adaptive management as "a fundamental Program concept" that is to be adopted and practiced by all program elements and throughout the Program as a whole:[5] "solutions must be guided by adaptive management. The Bay–Delta system is exceedingly complex, and it is subject to constant change as a result of factors as diverse as global warming and the introduction of exotic species. CALFED will need to adaptively manage the system as we learn from our actions and as conditions change" (CALFED 1999c, p. 15).

The Strategic Plan for Ecosystem Restoration (CALFED 1999d), published as a guiding document for ERP, closely follows the framework for adaptive management developed in the academic literature and discussed earlier in this chapter. Quoting liberally from work by Walters and Holling, the Strategic Plan lays out a step-by-step adaptive management methodology involving modeling, designing management interventions on the basis of these models, implementing and monitoring the interventions, and then redesigning the interventions in light of improved understanding.

Combined with the common programs have been competing proposals for new or expanded water storage and modifications of Delta conveyance, initially known as the Program Alternatives. Under Alternative 1, "existing system conveyance," Delta channels would have been maintained essentially in their current configuration. Alternative 2, "modified through-Delta conveyance," involved significant improvements to northern Delta channels accompanying the southern Delta improvements contemplated under the existing system conveyance alternative. Alternative 3, "dual Delta conveyance," formed around a combination of modified Delta channels and a new canal or pipeline connecting the Sacramento River in the northern Delta to the export facilities in the southern Delta of the State Water Project and the Central Valley Project.

The "Preferred Program Alternative," to be developed in stages, starts out as a modified through-Delta conveyance approach (Alternative 2). Depending on the results of that approach, CALFED might later adopt a dual Delta conveyance strategy, that is, construct an isolated conveyance facility (Alternative 3). Such a facility is vehemently opposed by most in the environmental community and was rejected in the early 1980s in a statewide referendum. For many, an isolated facility that takes water more directly from northern to southern California, thereby bypassing the Delta, would eventually put water, agriculture, recreational, and environmental resources in the Delta at increased risk. In response to the continuing polarization around such an alternative, even among its cooperating agencies, CALFED has decided that an isolated facility cannot be studied, approved, funded, and constructed within the first stage (7 years) of implementation.

How these program elements are to function and their projected costs are still being worked out. Clearly, the costs of the CALFED Bay–Delta Program will be shared by many entities, including user fees, federal appropriations, private–public partnerships and general obligation bonds. California Proposition 204 in 1996 provided more than $450 million for the CALFED Bay–Delta Program's environmental enhancement efforts. Federal authorization for an additional $430 million over the next three-year period has also been secured. Over its lifetime, the Program could cost as much as $10 billion, making it one of the most expensive projects of its kind in the world.

Interior Columbia Basin Ecosystem Management Project (ICBEMP) and Related Initiatives

The ICBEMP initiative was chartered in 1994 by the Forest Service (FS) and the Bureau of Land Management (BLM) in response to President Clinton's charge to "develop a scientifically sound and ecosystem-based strategy" for management of the Eastside forests in the region (Quigley et al. 1999, p. 273). The project has included a scientific assessment of the interior Columbia Basin within the United States and east of the Cascade crest and those portions of the Klamath and Great Basins in Oregon (hereafter, "the Basin"). The total assessment area includes approximately 227,000 square miles and portions of seven states (figure 2.2). The FS and BLM manage over half of the area (117,000 square miles) in 35 national forests and 17 BLM districts (for a brief description of FS and BLM, see table 2.3).

The lands in the Basin are highly varied. Mountain ranges in central Idaho and western Montana commonly have elevations of some 3,000 meters or more. The Basin covers extensive plateaus as well as deserts and plains. Grassland, shrubland, and woodlands are present in the region. Most of the area is drained by the Columbia River and its tributaries. The Basin covers about 8% of the U.S. land area and contains about 1.2% of that country's population. There are six metropolitan centers in the area, where

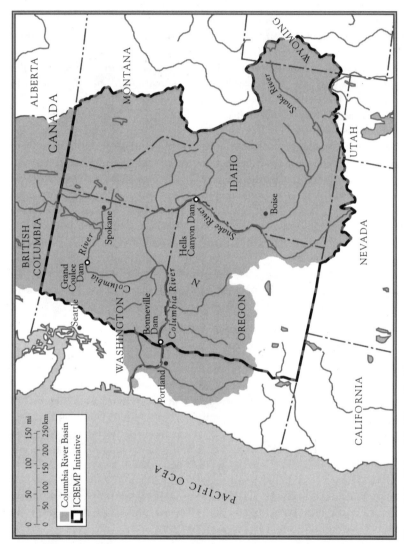

Figure 2.2 Columbia River Basin.

the majority of the people in the Basin live, though the area is primarily rural and agriculture is the dominant industry. There are also 22 recognized Native American tribes in the region with historical salmon fishing rights and claims. According to the assessment, it has been the six metropolitan counties that have been the engine of the Basin's economic growth, with higher rates of growth in total employment, total personal income, and non-farm labor income, and a greater ability to weather national recessions than other counties. On the whole, however, the Basin is a diversified economy of 1.5 million jobs, where per capita income is increasing faster than the U.S. rate.

The Pacific Northwest has changed dramatically since the descendants of Europeans began inhabiting the area 150 years ago. At the end of the 19th century, stream modifications were made to facilitate navigation on the Columbia River system. By the end of the 20th century, the waterway from the eastern city of Lewiston, Idaho to the Pacific Ocean had become a series of reservoirs. Today, a system of nearly 1,240 large dams and thousands of smaller dams provides an estimated 40% of the U.S. hydropower production, navigation and irrigation benefits, flood control, and recreational opportunities. Under recent rapid population growth, the area is being transformed from a long-standing resource-based economy into one that is more based on technology, transportation, and service sectors.

Conflict over the natural resources has marked the area for decades. Much public debate is framed in terms of commodity extraction versus resource protection. Concerns have grown about timber harvesting from public lands, declines in wildlife species in old forests, the decline of anadromous fish, threats to forest health (such as tree mortality caused by insects, disease, and wildfire), and rangeland health. The debate over fisheries has moved to center stage in the public debate, as anadromous fish populations (several species of salmon) have kept declining since the 1980s (see Lee 1993). Over 10% of the fish species found in the area are currently listed under the ESA as endangered or threatened, while others remain candidates for listing. The national forests and BLM districts in the area became involved when the ESA prompted new interim standards and guidelines and related lawsuits. The FS has been obligated to consult with the National Marine Fisheries Service (NMFS) on the implications of forest plans for listed fish species and their habitat.

FS and BLM are not the only agencies affected, nor are they the most affected. The Basin's many dams were seen as a, if not the, major threat to the listed salmon. The federal dams that generate most of the hydropower are operated by the Army Corps of Engineers (ACE) and the Bureau of Reclamation (BR). Their hydropower is sold by the Bonneville Power Administration (BPA), which is the region's main power supplier. BPA also sells surplus power to California and the southwestern United States. The crisis around salmon decline has had far-reaching effects on these organizations. Currently, BPA funds about 250 fish and wildlife projects a year, from repairing the spawning streams to studying fish diseases and controlling predators, totaling up to about one-fifth of the agency's operating budget. Projects for BPA funding are identified by the Northwest Power Planning Council's fish and wildlife program. The Northwest Power Planning Council (NPPC) is a four-state council formed by Idaho, Montana, Oregon and Washington to oversee electric power system planning and fish and wildlife recovery in the Columbia River Basin.

As discussed in the next chapter, these conflicts and problems put the NPPC and BPA at the heart of the paradox. Although these organizations are not part of the ICBEMP initiative, they are key to the process through which the Pacific Northwest tries to reconcile ecological rehabilitation and engineering reliability. For this reason, we have chosen to include their efforts in the case study along with our analysis of ICBEMP, so that terrestrial and aquatic ecosystems are treated together. More than in the other case studies, this region incorporates several major initiatives and activities to undertake more ecosystem management within a reliability context.

Faced with severe threats to salmon populations, FS and BLM also confront other endangered species issues as well as compelling concerns over riparian conditions, old-growth forests, and wildlife associated with the latter. The forests in the western part of the region have been at the heart of the public debate, and more recently the debate has widened to include protective measures for the eastern forests. Due to legal injunctions over endangered spotted owls and old-growth forest, timber harvest in the west has become virtually nonexistent. Now that logging in the east has also decreased, the gridlock in federal forest management seemed to become complete.

Such was the context for ICBEMP at its inception in 1994 (for more details, see Quigley et al. 1999). Goals of the project are given in table 2.6 and are considerably more ecosystem focused than those of the other projects (contrast with table 2.5).

Although less marked than in CALFED, ICBEMP has revolved around interagency collaboration. "A project leadership team, consisting of the

Table 2.6 The Six Primary Objectives of ICBEMP

- Maintain evolutionary and ecological processes.
- Manage with an understanding of multiple ecological domains and evolutionary timeframes.
- Maintain viable populations of native and desired nonnative species.
- Encourage social and economic resiliency.
- Manage for places with definable values: a "sense of place."
- Manage to maintain a mix of ecosystem goods, functions, and conditions that society wants.

Source: United States Department of Agriculture, Forest Service, and the United States Department of the Interior, Bureau of Land Management (1996, p. 30).

Science Integration Team (SIT) leader, two project managers, and the BLM coordinator, provided overall direction to the specific [project] teams organized around the principal products of the project assessment and EISs" (Quigley et al. 1999, p. 276). The SIT included representatives from the two leading agencies (FS and BLM) and from the Environmental Protection Agency (EPA), the U.S. Geological Survey (USGS), and the Bureau of Mines. Other cooperating agencies included NMFS and FWS. An Executive Steering Committee was established to provide direction and oversight of the project leadership team. Also, in the words of Quigley et al. (1999, p. 279), "the Executive Steering Committee consulted personally with each tribe in the project area, which required an individual approach for each *Idots* [especially] regarding water quality and salmon habitat."

The SIT has had primary responsibility for the scientific framework, assessments, and evaluation of alternatives for the project. The nature and scope of its duties has been a matter of some controversy, including congressional efforts to restrict aspects of the assessment. The SIT's final assessment, which took longer than expected and included many public meeting and discussions, "links landscape, aquatic, terrestrial, social, and economic characterizations to describe the biophysical and social systems" of the Basin study area. "The SIT has proposed that the assessment information be considered a part of a dynamic assessment that includes models, databases, and analysis updated through monitoring, inventory, and analysis processes" (Quigley et al. 1999, pp. 277, 283). The assessment was necessary and a primary product for the EIS/EIR. Once an EIS is issued, a Record of Decision is expected, which will be legally binding and thereby alter the planning and operations of the FS and BLM in the Basin.

On the basis of the assessment and as part of the EIS process, three alternatives have been identified and evaluated. Alternative S1 (roughly, a business-as-usual scenario) would basically continue existing management practices on FS and BLM lands. Alternative S2 was selected as the preferred alternative for the EIS process. All of the agency executives at the table (BLM, EPA, FWS, FS, NMFS) collaborated on this decision. Alternative S2 focuses on restoring and maintaining ecosystems across the project through

- restoring the health of the forests, rangelands, and aquatic systems in the project area;
- recovering imperiled species; avoiding future species listing; and
- providing a predictable level of goods and services from the public lands. (United States Department of Agriculture, Forest Service, and the United States Department of the Interior, Bureau of Land Management 2000a)

Alternative S3 is very similar to S2, in focusing on improving ecosystem functions and ensuring reliable services at the same time, though alternative S3 would accept more short-term risks to address the long-term risks more effectively.

The reason that ICBEMP's implementation costs are significantly lower than those of the other two initiatives (table 2.2) is that its implementation consists of changing the Forest Service and BLM land use plans. The implementation costs of the preferred alternative (S2) are calculated as the annual funds needed in addition to current budgets in order to implement S2 recommendations. Management direction from the project's Record of Decision (ROD) becomes part of the amended federal land-use plans and will guide decision making until replaced through subsequent amendment or revision. Within the ROD as the overall structure, it is expected that management decisions will change adaptively. Under the heading "Adaptive Management," the Supplemental Draft EIS states:

> The intent is for management direction to be modified if a site-specific situation is different than what was assumed during ICBEMP planning. ... Accelerated learning is intended to occur from formal research designed to test hypotheses of scientifically uncertain and/or controversial management issues, or to use field trials to test the usefulness of new strategies to achieve objectives. (United States Department of Agriculture, Forest Service, and the United States Department of the Interior, Bureau of Land Management 2000b, p. 16)

The overarching decisions taken by virtue of the ROD are expected to be in effect for roughly 10 to 15 years. Efforts have been made prior to the ROD to incorporate material from the assessment into the existing land-use plans of the two lead agencies. It was always intended, however, that the assessment would provide data and input for different planning and management activities in the future beyond FS and BLM land-use plans.

Comprehensive Everglades Restoration Plan

CALFED illustrates a consortium of state and federal agencies taking the lead in ecosystem rehabilitation and improving service reliability. In the Pacific Northwest case study, we see FS and BLM leading ICBEMP—while NPPC and BPA are involved with aquatic ecosystems. In Florida, a different mix of agencies has taken the lead, one federal (ACE) and one state (South Florida Water Management District, SFWMD).

The human hand has transformed the ecosystems studied in this book, but none more so, it seems, than the Everglades. Ogden (1999, p. 174) provides a good review of the recent changes in the Everglades. Researchers and other authors (including United States Army Corps of Engineers and South Florida Water Management District 1999a) found that the area of the original greater Everglades has been reduced by almost 50% due to the conversion of large portions to agriculture and later to urban land uses; the depths and distribution patterns of the water system in virtually all the remaining areas of the Everglades have been altered; approximately 70% less water flows through the Everglades of today than originally; of the seven major landscape features in the presettlement Everglades, three have been eliminated entirely; exotic species have been introduced into the Everglades, changing thousands of acres of wetlands; the population of alligators has declined by an astounding 98%, according to one estimate; the number of nesting wading birds has declined by 90–95%; and the number of threatened and endangered species was up to nearly 70 in 2000.

The greater Everglades ecosystem, called the south Florida ecosystem, stretches south from Orlando and includes the Kissimmee Valley, Lake Okeechobee, the remaining Everglades, and on to the waters of Florida Bay and the coral reefs. Between Lake Okeechobee and the Everglades National Park are the Everglades Agricultural Area and three Water Conservation Areas. This south Florida ecosystem is much larger than what most people see when they visit "the Everglades"—usually just the Everglades National Park. Overlaying the ecosystem—and connecting it to the coastal area of metropolises including Miami, Ft. Lauderdale and other urban centers—is an elaborate water management infrastructure, most notably the Central and Southern Florida Project built by ACE (figure 2.3).

Created through legislation in 1948, the Central and Southern Florida Project is managed by SFWMD and ACE. It encompasses an area of 18,000 square miles, including 1,000 miles of canals, 720 miles of levees, and almost 200 water control structures. The project's mandates are to ensure reliable water supply, flood protection, water management and related services to south Florida. Population in south Florida has risen from 500,000 in the 1950s to more than 6 million today. As a result, not only are the quality and quantity of the water that enters the ecosystem seriously degraded, but it is also widely accepted that there is not enough water for the people either. Water restrictions have been increasingly invoked in response to shortages. Shortages, however, have been as much a matter of timing as of quantity. The water management infrastructure currently shunts 1.7 billion gallons of freshwater into the ocean everyday, leaving the Everglades with too little water in the dry season and too much in the rainy season.

Legislation in 1992 and 1996 provided ACE with the authority to review the Central and Southern Florida Project. The Corps was asked to develop a Comprehensive Plan to restore and preserve the south Florida ecosystem, while enhancing water supplies and maintaining flood protection. Together with SFMWD, ACE undertook the Central and Southern Florida Project Comprehensive Review Study (known as the Restudy). The Restudy was achieved through the work of more than 100 ecologists, hydrologists, engineers and other professionals from more than 30 federal, state, tribal, and local agencies. The Restudy was the basis for the Central and Southern Florida Project Comprehensive Plan to restore and protect the south Florida ecosystem. The plan addresses the four fundamental problem areas of the quantity, quality, timing and distribution of water.

The agencies involved in the Restudy and the Plan were many and required considerable coordination. In addition to ACE and its non-federal cost-sharing partner SFWMD, the Restudy team involved other federal and

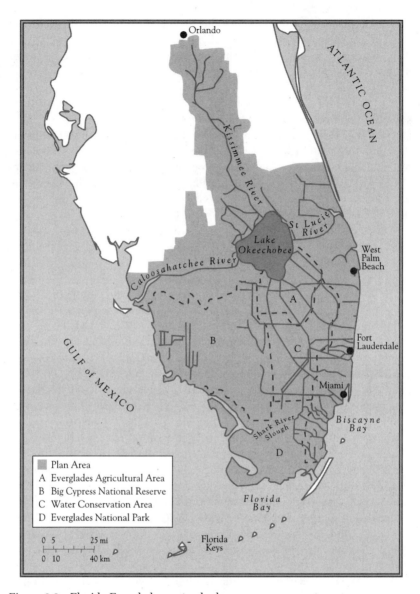

Figure 2.3   Florida Everglades watershed.

state agencies including the NPS, FWS, NMSF, USGS, EPA, Florida Game and Fresh Water Fish Commission, Florida Department of Environmental Protection, and the Florida Department of Agriculture and Consumer Services. As in the CALFED and ICBEMP initiatives, various Native American tribal groups were consulted during the Everglades Restudy. Local governments were involved from Miami-Dade, Broward, Palm Beach, Martin and Lee counties.

After a consultation and review process, the Restudy identified 40 preferred options organized around 13 "thematic concepts." Among the options, many explicitly concerned the restoration of the ecosystem and the reliable provision of water quality, quantity, timing and distribution. The thematic concepts also reflected the simultaneous improvement of ecosystem services and function, under headings such as "water supply and flood protection for urban and agricultural areas," "adequate water quality for ecosystem functioning," "invasive plant control," and "protection and restoration of coastal, estuarine, and marine ecosystems" (United States Army Corps of Engineers and South Florida Water Management District 1999b, pp. 6–9). More than the other two U.S. initiatives, the Plan for the Everglades revolves around increasing water storage as the way to address the problems of restoration and reliability.

The preferred options and thematic concepts were later screened and consolidated as part of generating alternative plans, from which the finalized Plan was generated. There were six alternative plans, developed by two teams, the Alternative Development Team and the Alternative Evaluation Team. The former was responsible for designing each alternative plan in response to the latter's evaluations of the previous plan iteration. Engineering and ecological models were important in the development and evaluation of the six alternatives. The iterative formulation and evaluation process refined and improved the model performance of subsequent alternative plans.

The finalized Plan has seven principal program components. Aquifer storage and recovery entails building over 300 wells to store water a thousand feet underground in the aquifer. Up to 1.6 billion gallons a day may be pumped into the underground storage zones. Stormwater treatment areas will be developed, with an estimated 56 square miles of human-made wetlands to be built. The areas will treat urban and agricultural runoff water before it is discharged into the greater ecosystem. Wastewater reuse is to be furthered through the construction of two advanced wastewater treatment plants. Millions of gallons of groundwater are lost each year as it seeps away from the Everglades towards the east coast, either as underground flow or through levees. The seepage management component of the Plan would reduce unwanted water loss through, for example, adding impervious barriers to the levees, and redirect this flow westward to the Water Conservation Areas and Everglades National Park. Removing barriers to the sheetflow in the Everglades is the sixth program component. "Sheetflow" is the slow and broad flow of water resulting from the low hydrological gradient and the resistance to flow caused by vegetation in the Everglades. More than 240 miles of project canals and internal levees within the Everglades will be removed to reestablish a more natural sheetflow of water. Finally, operation changes will have to be made in water delivery schedules to alleviate extreme fluctuations. Lake Okeechobee water levels will be modified to improve the health of the lake.

As in the CALFED and ICBEMP initiatives, the Comprehensive Everglades Restoration Plan is to be managed adaptively. In the words of the Plan summary (United States Army Corps of Engineers and South Florida Water Management District 1999a, p. 20), the "Plan is designed to allow project modifications that take advantage of what is learned from system responses, both expected and unexpected. Called adaptive assessment, and using a well-focused regional monitoring program, this approach will allow us to maximize environmental benefits while ensuring that restoration dollars are used wisely.... Independent scientific review is an integral part of this process."

The U.S. Congress recently approved the Plan for approximately $7.8 billion. In addition, the Plan will cost approximately $182 million a year to operate, maintain, and monitor. Of the recommended 68 projects, ten projects and the adaptive assessment program, totaling $1.1 billion, were recommended for initial Congressional authorization. Taken together over the more than 20 years needed to implement the Plan, the Plan's annual costs amount to just over $400 million. In general, the federal government will pay half the cost, and the state of Florida and SFWMD will pay the other half.

Implementation of the Comprehensive Everglades Restoration Plan, as with the CALFED Program and ICBEMP, is only now beginning. The initiatives are winding up their initial planning efforts and each has proposed detailed preferred alternatives. All of this is subject to revision, however, and the initiatives will be forced to rethink some or many of their decisions in light of the experiences to come.

Like so many of the ecosystem management efforts discussed earlier in this chapter's literature review, very little exists on the ground by way of ecosystem management in general and adaptive management in particular. Certainly nothing exists on the ground that has resolved the paradox. Why then look at the three initiatives? The initiatives would be of interest if simply because of the amounts of money involved. More important for our purposes, however, is that the process of undertaking assessments and generating alternatives—all driven by the EIS process—ensures that hard questions are addressed seriously. The ESA has ensured that the very real tradeoffs between species protection, ecosystem restoration, and water reliability are stark and unavoidable. The difficult tradeoffs between engineering reliability and ecological rehabilitation are now being faced and in some way initial priorities are being set. As our interviews show, these tradeoffs and priorities may well be glossed over and not fully recognized in the official project documentation, but managers face them nonetheless. We turn to the interviews now.

3 Adaptive Management in a High Reliability Context

Hard Problems, Partial Responses

The examples found at the beginning of this book are, to our minds, neither instances of a lack of societal commitment to saving the environment nor evidence of unreasonable demands for highly reliable services. If they were that, the obvious answer would then be to bite the bullet and take either the environment or the services more seriously. In our view, the examples really express the hard paradox of having to improve the environment while ensuring reliable services at the same time. Beyond specific examples, the strongest expressions of the paradox being taken seriously in terms of the budgets and stakes involved are those large-scale adaptive management initiatives proposed and undertaken in regions where they seem most difficult to implement; that is, where the reliable provision of services is a priority. Just what "reliability" is for the kinds of organizations we study is detailed in chapter 4. Here, we take a closer look at our case studies to see how the issues are articulated empirically.

Setting the Stage

The paradox is even enshrined in law. The mandate of the Pacific Northwest Electric Power Planning and Conservation Act of 1980, for example, is to "protect, mitigate and enhance fish and wildlife affected by the development, operation, and management of [power generation] facilities while assuring the Pacific Northwest an adequate, efficient, economical, and reliable water supply." But how to do this? Or, as one ecologist, Lance Gunderson (1999b, p. 27), phrased the paradox, "So how does one assess the unpredictable in order to manage the unmanageable?"

The answer usually given by ecologists and others is to "undertake adaptive management" (chapter 2). The decision maker learns by experimenting with the system or its elements, systematically and step-by-step, in order to develop greater insight into what is known and not known for managing ecosystem functions and services. Learning more on the ground about the system to be managed is imperative, especially given imprecisely defined terms such as "restore," "enhance," and "reliable." As the senior biologist planner at the Northwest Power Planning Council told us, the last clause of the Power Act "AERPS" (adequate, efficient, economical, and reliable power supply) "never has been quantified, so it is not very clear what it actually means." He is not alone. Berry et al. (1998, p. 62) conclude from their interviews in the Pactific Northwest, that "key concepts of ecosystem management (EM), adaptive management (AM), and ecosystem management research (EMR) are far from being consistently defined or generally understood, at least by those we interviewed. This might not create problems, except for the fact that our informants are responsible for the concepts' implementation."

Even if the overarching terms were clearly defined, the literature review and our interviews show that there would still be a lack of consensus, both over what is being "restored" (the species, its habitat, or the wider ecosystem) and what it is being restored to. The goal may be some ecosystem state predating human settlement or disturbance (i.e., the presettlement template or predisturbance regime), the way it was historically (say, in the 19th century), the way it once was for keystone species or processes, the way it could be (enhancing specific ecosystem complexity or variation, whether ever there or not), the way it should be (e.g., the "normative river" of idealized functions and services to which decision makers should all aspire), or the way the ecosystem *must* be in so far as decision makers must have clear goals and objectives if restoration is to work.[1] In the face of such imprecisions, and if simply because species, habitats, and ecosystems are extremely complex, adaptive learning becomes for many the *sine qua non* of good resource management.

In the case of the Columbia River Basin, the "opportunity to use adaptive management was created by the [Power] Act" (McConnaha and Paquet 1996, p. 414).[2] Whether in text or deed, adaptive management became a driver of ecosystem management initiatives as part of their authorizing documentation and implementation. "The delivery mechanism of ecosystem management is supposed to be adaptive management," contends the senior scientist at the Interior Columbia Basin Ecosystem Management Project (ICBEMP). Another interviewee says it more directly: "ecosystem management and adaptive management are the same. You can't have an ecosystem management that is non-adaptive." Chapter 2, moreover, underscored the prevailing equation of adaptive and ecosystem management in the literature.

So what are the problems in undertaking adaptive management in a high reliability context? While the mandates appears contradictory, this is the context that gave rise to the need for ecosystem management in the first place. High reliability resource requirements have frequently worked against ecosystems, as when massive waterworks such as dams and irrigation canals destroyed presettlement habitats in the name of ensuring reliable water and power supplies to cities and agriculture. The following introduces the problems, and responses they typically elicit, and argues for a new approach to address both.

Adaptive Management

Ecosystem management actually has *two* templates that have been difficult to realize in practice. One is the presettlement template, the other is adaptive management itself. It is nigh impossible to restore ecosystems back to what they were prior to the advent of human (typically European) settlement. Moreover, it is next to impossible to undertake adaptive management as it is typically recommended. Thus, ecosystem management has gotten itself into a double bind of two reinforcing unrealistic templates. But that is not the only problem in moving adaptive management from recommendation to reality.

Key to adaptive management is the experiment. "Adaptive management," according to Arrow et al. (1995, p. 95), "views regional development policy and management as 'experiments', where interventions at several scales are made to achieve understanding, to produce social or economic product and to identify options." A great deal of work on experiment-oriented adaptive management has taken place, with calls to extend its application (e.g., Holling 1978; Walters 1986, 1997; Walters and Holling 1990; Lee 1993; Gunderson et al. 1995; Haney and Power 1996; McLain and Lee 1996). In fact, we make a strong case in later chapters for adaptive management, though not for the cases and reasons usually recommended.

Many interrelated problems—none of which on its own is without a proposed response—have hampered the application of adaptive management:

- Ecological conceptual and quantitative models (hereafter "ecological models") on which to base experiments are rudimentary in nature, particularly for aquatic ecosystems. They are thus unable to provide much guidance on ways to improve ecosystems by preserving, restoring, or otherwise rehabilitating ecosystem processes and functions so as to better mimic a preexisting template (see Johnson 1999, on the difficulty of generating models for adaptive management).
- Where ecological models do exist, a great deal of field time is needed to estimate their parameters; where models do not exist, a great deal of study is needed to develop them. In either case, the empirical work is difficult to generalize beyond the area or site researched (e.g., Walters 1997).
- Adaptive management experiments have been designed in the field and on a large scale. However, in the absence of control groups and

replicable designs and in a world where testing the no-treatment option is unfeasible and where experiments must start with best management practices, many have not been (adequately) derived experimentally (e.g., Carpenter 1996).
- High costs are associated with experiments and the ecological risks of unintended impacts, especially when the lag time between experiment and discovery of associated error is long and the magnitude of the error large. In such cases, errors when discovered may be, for all practical purposes, irreversible. This is especially important as many proponents of adaptive management argue that real learning can only be achieved through large-scale—not scattered site—experimentation. While adaptive management is ideal when the potential error is small (or better yet non-existent) and the lag time short (or better yet immediate), ecosystem complexity ensures that ideal is unlikely to be achieved (e.g., Gunderson 1999a).
- Adaptive management as a structured process of learning-by-doing experimentation occurs in relatively unforgiving or even hostile political, organizational, and social environments that expect any such intervention to be of little or no risk, right the first time around, and with few, if any, opportunities to do that experiment again (e.g., Gunderson et al. 1995).
- There is a variable, in some cases, missing consensus over (i) how exactly to define adaptive management, (ii) the extent and intensity of organizational and political commitment to adaptive management, however defined, and (iii) how such defined adaptive management should actually be implemented across initiatives to improve ecosystems (e.g., Walters 1997).

For these reasons, it is not surprising that our interviews and review of ecosystem management initiatives found no example of adaptive management being implemented as proposed in the literature. What is often called adaptive management is for all intents and purposes a succession of one-off interventions: "experiments" that are non-replicable, large-scale, without control groups, based on the best information available, and undertaken in a world that is never one way only and where you rarely have another chance to do it over again. When asked to give examples of experiment-driven adaptive management, its advocates in CALFED responded by naming this or that intervention, then ending with the qualification, "but that wasn't a real experiment."

Shorn of its system-manipulation requirements, the experiment becomes like any other major management intervention. Nor is it is difficult to understand the many workaday reasons why the experiment has to be the intervention. As one CALFED interviewee frames it, agencies' reluctance to fund "further research" works against adaptive management. They assume that if you need more time to study the issue, then you obviously do not know what to do, so why give you the money in the first place?

Predictably then, "everyone does adaptive management, but no one really does adaptive management," says the senior Columbia River Basin biologist planner. "Are there any real adaptive management projects anywhere in the world?" he asks. He continues:

> A lot of the stuff we do is called adaptive management, but none of the experimental stuff that makes for real adaptive management ... [which] is tough to do. You can't experiment with ESA [Endangered Species Act] species: you have problems manipulating the hydrosystem to provide the necessary experimental conditions, there is a lack of agreement on what the question is, and there is a lack of prioritizing the importance of the different questions out there.

"Everyone is singing the tune that the only way to do ecosystem management is adaptive management," says one ecologist familiar with the Florida Everglades initiative. He feels, however, that adaptive management has become too narrowly defined. The ICBEMP senior scientist expresses the problem this way:

> The question of adaptive management has been a source of tension between scientists and managers. The managers have chosen a General Planning Model [with monitoring, assessment, decision making, and implementation components], but what [they] are doing is taking the GPM and saying that it is adaptive management. We say it doesn't include learning because it is missing large-scale experimentation and implementing it on purpose to learn.

"Adaptive management is on the lips of all [forest] managers," says the ICBEMP scientist, "but their actions, though improving, are still lagging behind their words." The actions and interventions may be called an experiment, a prototype, or a demonstration project, but whatever the phrase, the key characteristic is that, as with other major interventions, their designers understandably try to avoid error at all reasonable costs with the objective of getting it right—or mostly so—the first time around. Adaptive management is recommended as the way to learn what important uncertainties remain for better managing resources and ecosystems. In reality what decision makers want is enough certainty to enable their management of resources and ecosystems to continue, preferably better than before. We found interviewees quite reluctant to be publicly associated with any adaptive management that is anything remotely like trial-and-error learning, notwithstanding their understanding that experimental learning is frequently described this way.

Once the experiment is taken out of ecosystem management, it is difficult to see what is special about it. "We've been doing [ecosystem management] a lot over the years already, without calling it that," says the team leader of the Environmental Impact Statement (EIS) process related to ICBEMP. "A lot of [ecosystem management] is being done," says an ecologist familiar with overseas adaptive management efforts, "but it is not being called adaptive management." The temptation is to take any unexpected crisis as grounds for an "experiment." But as a biologist associated with CALFED's large ecosystem management initiative phrases it, crises are "reactive while adaptive management is supposed to be planned." While smaller experiments are more feasible, our interviewees make clear their preference for large-scale experiments in order to ensure that results can be generalized and scaled up. Problems with large-scale trials, however, are reminiscent of those faced by the large-scale social science experiments and demonstration projects of the 1960s and 1970s, which did a much better job of uncovering complexity than of reducing uncertainty (Nathan 1988).

The temptation is to dismiss adaptive management altogether. "Don't take it too seriously," says one informed CALFED insider of adaptive management. While the adaptive management framework is most developed in CALFED's Ecosystem Restoration Program, its wholesale extension to the other program components, including those with high reliability mandates, has been considerably less deliberated. CALFED's Long Term Levee Protection Plan devotes only one page to adaptive management, giving much more space to contingency planning, risk management strategies, and emergency response plans. Arguably, however, these latter elements would have been better treated as parts of the wider adaptive management strategy than as stand-alones.

To repeat, none of the preceding problems need be insurmountable. Recommendations for addressing each continue to be developed. The sense of intractability increases, however, when the problems occur and interact together, as they most notably do in human-dominated ecosystems in which population densities and intensified resource utilization place tough reliability requirements and constraints on ecosystem services and functions.

By now, it should be clear that the problems associated with adaptive management are not entirely derived from increased population densities and extraction. They also stem from the relative unavailability of adequate ecological models for the ecosystems and experiments based on such models; from other organizational demands competing with demands for protected, restored, or rehabilitated ecosystems; and from the increasingly stringent requirement that each ecosystem resource (e.g., water) reliably provide multiple services (e.g., for agricultural, urban, and environmental uses). We turn now to explore each of these five problems areas and the partial responses they have elicited to date.

Hard Problems

Increased Human Population Densities and Resource Extraction

"All this talk about ecosystem management doesn't mean a damn thing until we can somehow control our population growth," says a U.S. Fish and Wildlife Service (FWS) manager in Grumbine (1997, p. 42). The most visible effect of human population growth and extraction is the massive footprint they leave across the landscape (Vitousek et al. 1997). Ecosystems have been transformed irrevocably. Yet the relatively invisible effects of population growth and extraction have left the most lasting stamp and it is here where we must first turn our attention.

System Interconnections and Complexity Rapid population growth and human extraction have left the people of this planet with highly interconnected systems—economic, social, political, cultural, organizational, and ecological. They are complex and dynamically changing to such an extent that many management interventions do not succeed as planned or they actually lead to unintended consequences counter to the original management objectives. This phenomenon has been widely commented on (most popularly as the "revenge effect" in Tenner 1996), and ecosystem management certainly has not been immune to its effects.

Examples of the unintended consequences phenomenon are widely reported (see chapter 1). Take the 1999 delta smelt crisis in California. A handful of the ESA-listed smelt in a sample is enough to close a cross-channel gate, which in turn reduces water quality thereby breaching urban water standards and lowering water exports to agriculture by half. Another crisis: releasing dam water for salmon passage means less electricity for the Bonneville Power Administration's grid. Combined with hot weather, sagging power lines, and a broken backup generator, this leads to the electricity grid becoming unstable, causing the shutdown of the California–Oregon intertie and a massive blackout in California. The list can easily be lengthened.

The chief implication of such examples is that any intervention, including adaptive management, must be seen as part and parcel of a highly interconnected set of systems, where the intervention, planned or not, could have widespread but unknown effects. Interventions thus pose problems that are rarely resolved directly. Because systems are interlinked and complex, their responses to change are as complex (e.g., Melillo 1998, p. 183). One of our ecologist interviewees reports: "It's a lot more complicated than restoring variability because it gets to the issue that these systems are themselves more complicated. If you try to do restoration straightforwardly you are likely to fail, because of unintended effects due [to what he called the] accumulated nasties or vampires in the basement. ... When you go about reestablishing variability the vampires get out."

"The deliberate manipulation or management of ecosystems, therefore, will almost certainly involve some untoward surprises," conclude Callicott et al. (1999, p. 28). They add, "Adaptive management assumes that such surprises will occur as a matter of course." These inevitable surprises and their potential for unintended consequences are indeed a primary reason used to justify the need for adaptive management. Why? Because it is only through gradual, experiment-based learning by doing, or so it is argued, that decision makers can find their way through this dense tangle of causality to better management.

Table 3.1 Suitability of Trial-and-Error Learning

		Lag time in discovery of error	
		Short	Long
Magnitude of error	Small	Best	Better
	Large	Bad	Worst

Even here there is risk, which is why, according to one CALFED interviewee, they never use the phrase "trial-and-error learning" in discussions about adaptive management. Think of trial-and-error learning, or experimental learning more generally, along two dimensions (T.R. La Porte 1999, pers. comm.): (i) the short or long lag time between making an error and discovery of the error, and (ii) the magnitude of that discovered error, either large or small. The literature on adaptive management refers more to surprise than to error, where the former occurs with or without human intention and the latter requires human intention. As will become clear in a moment, our definition of error includes surprise, because the starting point is management. Table 3.1 sets out the four possible situations. Holding the scale constant (a condition we relax in chapter 4), adaptive management is ideal for situations where the lag time between intervention and impact is short and where the impact is not overwhelming. Adaptive management in the other three cells is more problematic. Indeed, when the lag time between the intervention and its effects is long, and the effects in turn are unpredicted, large, and with many remaining unknowns, decision makers really cannot call the intervention an experiment, no matter how capacious the definition of the latter.

The ecosystem is always complex and changing, something especially true for the aquatic ecosystems under study in this book. Thus, the large-scale experiment—or for that matter *any large-scale management intervention, including those by engineers*—will set in motion a series of events and impacts that entail unknown and possibly massive effects later on. Errors resulting from an ecosystem-wide experiment or systemwide engineering intervention may themselves be irreversible in a system that cannot be controlled and in which decision makers learn only years afterwards that what took place was indeed a mistake. The potential threat is not confined to the environment or the high reliability management of levees, power generation, water supplies and water quality—programs whose entire mandate is to avoid catastrophic or irreversible error. Restoration interventions also entail ecological risks. As one CALFED interviewee points out, these include the possibility of flooding, degraded water quality, and invasion of nonnative species around the experimental field sites (though, of course, non-intervention also entails certain ecological risks). Biologist D.W. Schindler concludes: "Sadly, these [adaptive management] 'experiments' are increasingly done on landscape scales where any but the smallest mistakes cannot be corrected, where the cumulative effects of several simultaneous insults must be interpreted and where political and social resistance prevent timely responses" (1996, p. 18).

Under conditions of cumulative, catastrophic, and irreversible errors, the prime management question can all too quickly become, "Who is criminally liable for the bad results of ecosystem management?" The answer increasingly is the high reliability organizations responsible for power, water, and the levees, as they are also mandated to practice ecosystem management.

Yet liable or not, decision makers feel increasing pressure to undertake ecosystem management that is adaptive in the best sense of the word, precisely because of the complex, interconnected task environment posed in all the table 3.1 situations. Since in this view anything is possible, everything must be at risk. If everything is connected, then everything could be equally important, and everything is potentially urgent. The short run becomes telescoped into what one our interviewees terms the "short-short term," where, in the words of another interviewee, "species under threat are treated as emergency room casualties." The long run is foreshortened as well. When the short-short term is daily or even hourly, it is not surprising that people think of the long term as a year or more out, as we found in Bonneville Power Administration's power scheduling and planning. Consequently, responsive ecosystem management becomes all the more urgent at the same time as it becomes riskier to undertake.

If such conclusions are not sufficiently sobering, the tightly coupled, interactively complex nature of the systems arising out of substantial population growth and extraction has two other highly consequential implications for ecosystem management. One has already been introduced, namely, ecosystems have been irrevocably altered, making realization of the presettlement template remoter than ever. The other is the extinction or decline of species and biodiversity, leading in the United States to the Endangered Species Act and its continued cascade of effects within and beyond the nation. These two implications are discussed briefly, as the topics are more fully developed in subsequent chapters.

Presettlement Template It is bad enough that species are being driven to extinction, but the other long-term casualty of population growth and extraction also deserves mention: the presettlement template. The view prevailing in the literature is that there is no returning to the way things were. Even our very disparate interviewees are unanimous on this point. "No one would think of trying to recreate the old flows of the Colorado River," says an ecologist. "We can't turn the Columbia back to what it was before, but we should be asking what can we do to improve things better than what they are now," concludes an official at the Bonneville Power Administration (BPA). "It doesn't make sense to blindly move the historic template into management objectives," says the ICBEMP senior scientist. "I've always doubted the usefulness of the presettlement template," says another ecologist, adding that the template was "a dynamic environment, not static or one point in time." In the words of an Everglades planner,

> [You] have to face the fact that this [proposed Comprehensive Everglades Restoration Plan] isn't the romantic notion of taking all the levees out on the argument that that is what the original system looked like. It would be a more natural system, but it wouldn't look like the Everglades. ... The [Everglades] is so fundamentally altered that it is on life support at the margins. It's like Disney World. You go into the attraction, and you see the pirates and it looks all very real, and then you look in the basement and see all the technology needed to be in place to maintain the illusion upstairs. That is what all the water [pumping] facilities are doing around the Everglades, it is creating the artificial experience that there is a natural system called the Everglades. It is a totally managed system, and we can't go back. We'll be managing it forever.

These views are not isolated to our case studies. They are found at other sites as well (e.g., Rhoads et al. 1999). Even if it were possible to restore an ecosystem back to presettlement conditions, there is no obvious reason why this is the state one would want the ecosystem to bounce back to when it suffers shocks in the future. In the words of the well-known zoologist, Gordon Orians (1996, p. 26), "ecosystems are naturally highly variable, and a return to a previous state is not necessarily an appropriate or desired outcome in all or even the majority of cases." Restoration and rehabilitation, as we saw in chapter 2, depend on the goals and objectives of ecosystem management. The absence of a presettlement template thus does not mean that ecosystem management is bereft of a template—or better yet, templates. We return to this point below under the responses to problems of population growth and extraction.

Endangered Species The cost of population growth and extraction is not just endangered species, it is also the Endangered Species Act. Much has been written about the iron mask of the ESA and related legislation—or perhaps not so much the law itself, as the narrow way in which it is currently implemented (chapter 1). Simply put—if that is the right phrase—the implementation of the ESA has changed policy, institutions, and their incentive structures so as to reduce dramatically the flexibility of both adaptive management and high reliability management. In making it harder to destroy specific species, the ESA has made it even harder to adaptively manage their habitat and ecosystems, at least in ways that optimize the use of the high reliability infrastructure already in place to provide water and other resources needed to restore, rehabilitate, and sustain habitats. For example, the ESA has halted active adaptive management in the Columbia River Basin, according to informed observers (Gunderson 1999b, p. 35). Nothing is made simple or flexible by the ESA, in the name of bringing species back from the cusp of extinction. It is also true, however, that the threat of the ESA was and is still the original motivator behind the search and resources for more adaptive approaches.

The ESA's species-by-species approach is often contrasted with the multi-species, habitat orientation of ecosystem management. The more compelling management dilemma is that the ESA is a fixed, comparatively inflexible management strategy. Yet the consensus is that the resource being managed requires management flexibility—particularly when the resource itself is mobile, such as fish. ESA inflexibility is altogether unnatural. With the ESA, you end up "making the Everglades into a zoo for single species," says a senior Everglades planner.

One must appreciate that *both* adaptive management and high reliability require flexibility in order to be effective. The ESA has changed the calculus of flexible service reliability. When speaking about BPA's operation of dams, a high-ranking engineer described the change thus "The ESA has greatly reduced the flexibility of the system. Flexibility means how power and floods would be managed differently without considering fish needs. ... In the 80s [we] said we are never going to spend [so] many millions on fish. Now [we] are doing it."

Salmon recovery has diminished the reliability of the system, one senior BPA official asserts, and this has come about through the ESA. As detailed in chapter 5, the Act has transformed the optimization problem for service reliability. In the words of a BPA engineer:

> In the past, because it was flood control and power, you were able to draw down [the resrvoir] and not worry about the fish. Now you must meet flood control and fish with power being secondary... Long-term planning in the projects no longer has any flexibility because of the many constraints. ... In the US almost all the flexibility is allocated for fish and flood control. Power has become an incidental benefit.

The effect of the ESA on adaptive management has been just as direct. According to an ecologist familiar with the Everglades and other ecosystem management initiatives, "If you don't know how to save [the species] or if what you are doing isn't working, then you have to try new things like experiments. On the other hand, you are not allowed to fail under the current interpretation of the ESA. We can't do anything until we know it won't cause harm. ... This is butt-stupid."

This necessity to take risks drives the search for flexibility in adaptive management, says the ecologist. In his view, two things are important for successful adaptive management: one is finding flexibility in the resource to be managed (such as identifying surplus water); the other is finding social flexibility, that is, the willingness of stakeholders to experiment and manage adaptively.[3] Since you can not experiment with endangered species, he says, that resource does not have flexibility in the first sense, while the lack of flexibility in the second sense is what killed adaptive management in the Everglades, in his view. "In a nutshell, if there is no resilience in the ecological system, nor flexibility among stakeholders in the coupled social system, then one simply cannot manage adaptively" (Gunderson 1999a). One of the most oft-repeated phrases in the adaptive management literature is the need and search "for more flexible, adaptive institutions and policies" (e.g., Holling 1995, p. 9; Carpenter et al. 1999).

Availability of Adequate Models and Their Alternatives

While the ecological models used in ecosystem management are improving (e.g., Carpenter 1999), a basic fact of life for adaptive management is that the ecological models it relies on are typically less adequate than those used in the provision of reliable services.[4] A variety of hydrologic flow and other simulation models (hereafter "engineering models") utilized in the state and federal water and power projects are much more fully developed and utilized than, for example, fish biology or population viability models. One reason is that engineering models have a much longer history than do ecological models, and, at least as important, water is easier to model than fish. The Bonneville Power Administration's oldest hydroregulation model dates back to the late 1950s (Bonneville Power Administration, Army Corps of Engineers, and Bureau of Reclamation 1992). In the words of the senior Everglades planner, "reliability is fairly well defined, performance measures are pretty clear, causal models [are] much better [for] reliability, [which is] not the case when it comes to the notion of 'if you wet it, it will grow.'" The lead ecologist in the South Florida Water Management District (SFWMD) says the hydrology models were very predictive and accurate, but not so for the natural system model (NSM), which "takes out levees et cetera and sees what would happen as if it were a hundred years ago. We feel it is the best estimate and it is under refinement as a predrainage template, but we can't use it [for fine grain analysis]."

"We have great hydrological and biogeochemical models," in the view of a well-known ecologist, "but no one has put them together and figur[ed] out how to link them. That's what I want to do. For [some] this could be done, [but] not for [others] because the models aren't there." All of this is important in her view because

> [w]hen you talk about adaptive management, you recognize that if you have good models, then you can do adaptive management, but if you don't have the models or an understanding of the system, then adaptive management isn't going to work. An alternative would be to do so much monitoring that you can model the causes and mechanisms empirically. But if you have good models you don't need so much monitoring.

Incompletely specified ecological models can have substantial implications not only for adaptive management interventions but also for the high reliability requirements based on those interventions. An example of conflicting interpretations of the impacts on seagrass of freshwater flows into Florida Bay was given by an ecologist familiar with the Everglades:

One hypothesis was that decreased freshwater flows created hypersalinity, which killed off the seagrass. Another was that nutrient flows into the Bay were causing the problem. A third was that the removal of keystone species that fed on biomass had an effect. A fourth was that the [seagrass] die-off was due to lack of hurricanes and other disturbances. Hard to say which one is correct and clearly not as easily resolvable as in water flow [i.e., reliability] issues. How in Florida Bay do you sort among these competing alternatives? If you negotiated with water reliability people around the need for freshwater and then found it was the turtle [that was the keystone species issue], then you would have spent your credit with the water people unnecessarily.

The "credit" that regulatory ecologists and biologists have with high reliability engineers and line operators is at the heart of the paradox. Ecologists may (again writ large) be perceived as ultimately precipitating cascading crises like that over delta smelt or the Columbia River Basin blackout of California power users.

Not only is there a dearth of adequate ecological models in adaptive management, but the ones that we have are often conflict with each other. Dueling models are a common complaint among our interviewees. "It's my model against your model," says one ecologist. "There've been dueling models in [the Columbia River Basin] for 10 years at least. . . . You can find a model to support your own policy conclusion," concludes the senior BPA official. "The Army Corps of Engineers . . . have their own assumptions and models; we [in the Forest Service] make our own assumptions and models. . . . No one has a fully consistent model that links models and assumptions," observes an ICBEMP scientist. Part of the conflict, of course, is that the different models ask different questions, only some of which may address the key uncertainties on which adaptive management is focused.

While there are models that try to link different models or layers of data (notably the ATLSS modeling approach for the Everglades), our interviewees' repeated observation is that each of the important regulatory agencies—National Marine Fisheries Service (NMFS), Army Corps of Engineers (ACE), Fish and Wildlife Service (FWS), Forest Service (FS), Bureau of Land Management (BLM), National Park Service (NPS), Bureau of Reclamation (BR) and others—had or wanted its own model(s), often for the same resource or species. "We have different models from the different agencies," notes the EIS team leader, "and each is somewhat different regarding what the effect is on the fish. . . . Even among the NMFS models there were differences in the assumptions about what makes for fish survival." Also, if each endangered species gets its own model, the conflict between models moves to a new level. "Sparrow is only one of the listed species," says the senior Everglades planner, "and what you want to do is avoid dueling species that are listed."

While there have been notable successes in modeling (described below), there is a sense that overall performance and results have been disappointing, when not irrelevant. "The problem with the Everglades," in the view of an ecologist, "is that there has always been a lot of money to spend on it, including models. ... This has produced modeling without a lot of results." An important part of the problem is the science underlying the models—that is, the perceived lack of adequate science. According to the BPA official, "We are working with NMFS, FWS, ACE, and BR, along with state agencies and tribes to implement the ESA. It has been a struggle going back to the first listed species, and [we] have attempted to use the best available science to set policy. That said, the information that is available is scant, given the surprising number of dollars spent."

The engineering models are not without their own problems, including dueling models and lack of data at levels of finer resolution. These difficulties, too, have implications for ecosystem management. The senior planner for the Everglades Comprehensive Plan notes that "flood control impacts are much harder to model at the landscape level, because they depend on very specific, local circumstances." This is all the more important because "you can't have real restoration without integrating it with flood control measures," in the view of an ecologist familiar with the CALFED exercise—and restoration also takes place at finer levels of resolution.

Another problem with the ecological models is their tradeoff between accuracy and user-friendliness. By making a model more accurate, modelers currently make it more complex, which renders it less transparent to users. "Our models are not user-friendly," says the ATLSS (Across Trophic Level System Simulation); modeler, "a lot of accuracy of the [ATLSS] model is bought by making it more complex; for example, we [presently] can't put the model on a PC." Some argue that a focus on model accuracy is overly narrow for other reasons. Insight, not accuracy, is the goal of good ecological modeling. "More detail is not the way to go in modeling," says an ecologist familiar with the Everglades, but "rather use models as another way to talk to one another." Models, he and many others feel, are at their most useful when helping us to ask better questions.

Nor does adding more complexity to the models necessarily help improve experiment design and hypotheses testing in adaptive management. The systems being modeled are empirically inseparable from the tightly interconnected systems in which the experiments take place. Replication of the experiment becomes problematic in the extreme. "Anyway, what is the replicate for modeling the Interior Columbia Basin?" asks the senior scientist in the Interior Columbia Basin Ecosystem Management Project (ICBEMP), "Is it the Amazon? The Colorado? The Mississippi?" How do you learn from a sample of one? (see also March et al. 1991).

The tangled system interconnections also make monitoring problematic. Where the models are missing, the next best thing is the monitoring and assessment noted earlier. Regrettably, in the absence of causal theories about the behavior of interest—that is, in the absence of ecological models—it is difficult to identify what indicators should be monitored. The decision maker ends up trying to monitor everything from a baseline that seeks to assess as much as possible. The Everglades ecologist agreed that a causal theory is important for identifying indicators, but in the meantime, "let's measure all we can," he advised. And that is just what many of the ecosystem management initiatives try to do. In early 1999, there were more than 600 different monitoring programs and units relevant to the CALFED Program. More than half pertained to environmental restoration. We heard repeated reports of monitoring programs that generated masses of data, little of which was being used. Of course, many of these programs were originally created for reasons other than ecosystem management.

Pressure to develop clear goals and objectives is strong in ecosystem management under these circumstances (chapter 2). Explicit goals replace not only the presettlement template as the guide for both restoration and experimentation in adaptive management; they also entail choices about the indicators to be followed. When asked how she would measure success in her proposed ecosystem management research project involving an interlinked pasture–forest system, an ecologist says,

> [I] don't know. You state your goals, you have to have some ability to develop indicators. ... The harder question is how do you know if you don't meet those indicators? Do you know why when you don't meet your goals? Can we explain why it didn't work? Do we have the models? How do we tell if the failure was due to factors internal to the ecosystem or due to linkages between systems? That's where understanding the links comes in. If the goal of increasing biodiversity is not being met, does it have something to do with the management of the pastures? We worry about how we can understand this. Can we really view the system as a system? Then if something is going wrong, how do we know if it's internal to that [subsystem] or whether it is because of the linked interactions? This has to do with the resilience of the whole system. Is crashing productivity a response to an external change like climate change? And if so, is it a direct response to, for example, different precipitation patterns? Or is it a response to a change in another subsystem, which has ... moved over a threshold?

She is not alone in her queries. When asked how he knew his agency's restoration program was effective, the senior Columbia River Basin biologist planner said, "We don't. People would like us to be real simple. Restoring 10 kilometers of habitat produces so many fish: that can't be done. You can only say something in the aggregate. The only way to tell [you are effective] is in the aggregate."

However, even clear goals and associated indicators may be impossible. Why? Because of precisely the same complexity that works against adequate ecological modeling. "In hindsight, initial demands for definitive objectives have been somewhat naïve because of the inherent complexity and conflicting nature of biological, social, economic, and administrative goals," write Johnson and Williams (1999) about an adaptive management harvest program run by the U.S. Fish and Wildlife Service. In the absence of clear goals, the call is for clear, researchable questions to drive adaptive management, as when Lee (1999) argues that "Adaptive management should be used only after disputing parties have agreed to an agenda of questions to be answered using the adaptive approach; this is not how the approach has been used." Models, of course, are one exceptionally good way to develop such questions, and it is the adequacy of these models that is in question. Indeed, one of the hopes of adaptive management has been to produce new models to address management questions (G. Peterson 2000, pers. comm.).

The call for a whole-system view or strategic focus when undertaking ecosystem management is just as strong as the drive for clear goals. Yet, the whole-system view or strategic focus is by definition at the large scale and over the long term. The level of analysis must then perforce take into account the causally complex set of systemwide interconnections, regardless of whether the ecosystem management involves adaptive management. When the large-scale and long-term perspectives are imperative and the causal relationships unclear or indeterminate, it becomes especially difficult to conclude that, for example, an ESA-listed species is actually recovering. One CALFED biologist showed us a frequency table for a listed fish species over the last 20 or so years. He poked his finger at it asking, "Is that recovery? Is that really recovery?" The table had frequency bars spiking up and down in no discernible pattern. Though the situation seems to have improved, was it because these were exceptionally wet years? Or was something else going on? One CALFED ecologist was worried about measuring the effectiveness of the Environmental Water Account (EWA), if it were actually implemented. The effectiveness of the proposed EWA had been measured only in a gaming strategy where the historical timeline served as a benchmark. The performance of EWA could thus be measured against the historical benchmark of what had actually happened without the EWA. For the future, however, no such benchmark exists. So the decision maker does not know whether the EWA actually is effective until long after the fact.

The CALFED initiative provides a good case study on how all the difficulties with modeling come together in ecosystem management (these are detailed in the Appendix).

Increasingly Competing Organizational Demands

For those unfamiliar with the American regulatory setting, the many county, state, and federal agencies involved in the push and pull of land and resource management across that country must seem both bizarre and breathtaking. When it comes to ecosystem management, alphabet soup—EPA, NMFS, BLM, FS, DWR, SFWMD, MWD—surrounded by short servings of the Bureau, Corps, Park Service, Fish and Wildlife, and Fish and Game, all spiced with EISs, EIRs, and RODs and served on the table that the ESA built, make for warring flavors, indigestion, and worse. Organizational and institutional pathologies are common: interim 18-month guidelines for the Columbia River Basin are in force for four years and counting; agriculturalists who witnessed the start of a south Florida water project still feel they have a right to that water, even though urban property owners pay the bulk of the taxes for its operation.

A leitmotiv in the literature and our interviews is the multiple, negative impacts of agency and stakeholder fragmentation, turf battles, and preoccupation, along with narrow or conflicting bureaucratic mandates on ecosystem management, including adaptive management. "Everything is one giant collage with everyone pushing their own agenda," says one of the Berry et al. (1998, p. 61) Northwest Pacific interviewees. Problems arise largely from fixed or inflexible management strategies in a context that many feel necessitates flexibility. The action agencies in the Bonneville Power Administration's technical management team, its major short-term planning unit, include the BPA, Bureau of Reclamation, Fish and Wildlife Service, and National Marine Fisheries Service. "How does it work?" we asked one of its members. "Poorly," was the reply: "Fish agencies come in with fixed positions, not willing to compromise or look for solutions. We can't get to the tradeoffs, they just state what they want. They want their cake and to eat it too. They can't differentiate, [they] want instantaneous water and also longer term water." "There are separate processes defined for every decision," says the ICBEMP senior scientist, "There is not a forum for resolving the full complex of the system and there is not one planned. We prefer a morass."

Organizational fragmentation aids and is abetted by separate fields of expertise, disciplines, and professions. Different professionals see the same thing in different ways. One CALFED scientist says, "Biologists don't believe in reliability," to which a line operator responsible for ensuring the reliability of California's State Water Project responded, "We can't believe in anything else." According to the Columbia River Basin biologist planner, "fish people think about abundance [in species numbers], while wildlife people look at habitat conditions on the assumption that when you build it, they will come." The differing professional orientations, whether toward species recovery, habitat restoration, or high reliability, have profound effects for restoring functions and improving the reliability of services, as we will see in chapter 6's discussion of blind spots. It turns out that just as there are professional differences, there are also surprising overlaps and opportunities for meeting the twofold goal.

A task environment that accentuates the differences among professions and agencies makes adaptive management more difficult. "The NEPA [National Environmental Protection Act] amendment process ... could take months or years if [the change in a management plan] is controversial," according to the head of the ICBEMP Environmental Impact Statement process. So, she says, adaptive management is difficult because it is hard to be adaptive in such an environment: "But there is a limit in terms of how congruent the [agency] mandates are. It has worked better with FWS [Fish and Wildlife Service] than NMFS [National Marine Fisheries Service] because FWS's mandate is broader. They also deal with plants, aquatics, land management.... At some point you have to agree to disagree because of different mandates and philosophies."

The same mandates, as we have seen, have a negative impact on high reliability management, particularly with respect to the impact of the ESA on water supply agencies. An eminent concern of the high reliability water agencies is drought proofing, as the senior Everglades planner puts it. What has happened, however, as organizational demands on the agencies increasingly compete, is that, in addition to water droughts, there are "regulatory droughts" and "ESA droughts," says a CALFED biologist.

Rising Demand for Multiple Services from Ecosystems

A frequent complaint in the organizational literature is that a bureaucracy's goals and objectives are scarcely ever defined with the degree of clarity that many would like. Pressure for explicit goals, objectives, and priorities in ecosystem management has already been noted, as these goald and objectives often lack the clarity that decision makers feel they need. That said, adaptive management problems also arise because the multiple goals and objectives of the Everglades, Columbia River Basin, and CALFED initiatives are clear and clearly in conflict, and nowhere more so than with respect to the paradox of how decision makers can reliably maintain multiple ecosystem services from increasingly unpredictable ecosystems. Reliability in service provision is a high priority in each of the initiatives. Indeed, it sets the context for such initiatives in three ways, each expressing the hard paradox: (1) the historical impacts of high service reliability have harmed the environment (more in a moment), thus necessitating the initiatives; (2) the response to these impacts has been to increase the funds and political stakes involved in improving the environment, thus creating the initiatives;[5] and (3) the ecosystem management initiatives themselves are bounded by high reliability management requirements in the process of meeting their goals. Three of the four goals of the CALFED Program, for instance, are directed at improving the reliability of levees, water supply, and water quality, while the fourth goal of "ecosystem quality" includes the reliable and safe restoration of degraded ecosystems.

Reliability, again, has been the great enemy of ecosystems. According to the Columbia River Basin planner biologist,

> In the past we have always tried to engineer the fish. So fish passage measures in the Columbia River, such as screens, for example, were always meant to be efficient in terms of power generation, namely, these measures were limited to a certain window in time; the rest of the year power would come first. And of course, over the years the window got smaller and smaller. You are slicing off parts of variable biological processes... slicing out biological diversity this way. Another problem is that we have "cleaned up" the stream, put riprap on everything. In other words, we have turned it into a drainage canal instead of

a river, turned it into a hydroelectric machine. We have destroyed biological variety in the process. ... [Ecosystem management] is not like fixing a car. That means you cannot think that we can really understand the system if only we look close enough. It is not a car that you can get to know in detail and that has the same repair instructions each time.

"Ecosystems," Grumbine (1997, p. 45) reiterates, "are not machines."[6]

Machines or not, decision makers are still left with the paradox. The same water has to save both resident fish in the dams and the anadromous fish downstream; the same water project has to serve urban, agricultural, and environmental users now; and the same agency has to restore habitat and make sure that power and water are delivered. Nowhere are the reliability demands more urgent than at the interface, where the services people expect from the ecosystem are being redefined the fastest and most intensely. According to the ICBEMP's lead scientist, "the biggest conflict is where the human interface is ... where more and more urban areas want to use more rural and wildlife resources and are moving into these areas."

Yet such reliable services are frequently forced into zero-sum choices. The Bonneville Power Administration is committed not to expand its system in the face of new demand and declining electricity load growth. In this way, the power demands of its chief customers and the nonpower requirements placed on the BPA are rising and increasingly in conflict with each other. A senior BPA official reports,

> in terms of allocating the [power] inventory, there are also nonpower requirements that we manage in the river system, which have the effect of reducing the firm inventory available to [our] customers. Nonpower requirements include water for salmon recovery (which is by far the biggest user in this category), flood control, irrigation, tribal needs (largely under salmon recovery), and possible navigation requirements. ... Each of the nonpower requirements wants its own water ... [and] salmon recovery has had an impact on our three [chief] customers.

A senior BPA engineer is more specific. In his words, there is "no end to the complexity" of trying to manage the conflicting water demands for the handful of listed or near-listed species under the ESA. A senior manager in the SFWMD operations office worried that the windows of opportunity in which his office could enhance the water supply to one listed species might conflict with the water supply needs of other listed species.

For these and other reasons, reliability is not only instrumental in effective ecosystem management initiatives, it is a deal-maker or breaker in the view of many stakeholders. The promise of storage facilities and other measures to double the water supply in south Florida is clearly the fulcrum on which the seesaw of stakeholder support is balanced for the Everglades Comprehensive Plan (chapter 5). Conversely, the lack of similar assurances can be a key factor in an initiative's problems. In a 1998 policy position paper for CALFED, the Metropolitan Water District of Southern California (MWDSC), the largest of its kind in the United States, was forthright about the priority of reliability and stakes involved:

> Metropolitan has become increasingly troubled regarding the recent shift by CALFED away from selecting a technically superior alternative. ... Instead, CALFED has defined a phased decision-making approach, which to date includes no clear commitment to provide water quality and reliability benefits for urban California. ... Funding and implementation of ecosystem restoration, in-stream quality improvements, and levee repairs should not proceed unless commensurate funding and implementation of safe drinking water quality, salinity control, and supply reliability also occurs. ... Finally, unless supply reliability is improved, Southern California's $500 billion economy, quality of life, and environment will be threatened.

As one might expect, a great deal of Metropolitan's own activities are directed to ensuring or enhancing the reliability of its water supply. Its own storage reservoirs are expressly meant to cover contingencies such as earthquakes and drought, providing a source of redundancy in the high reliability system for the Los Angeles water supply (e.g., its eastside reservoir when at capacity has a three- to six-month water supply at current levels of demand). Assuring reliability of water supply and quality is paramount among other stakeholders as well. As one stakeholder representative puts it, "The [CALFED] water quality objective is reliability."

Water reliability has the same (or greater) prominence and priority as ecosystem restoration in the ecosystem management initiatives reviewed or visited for this book. CALFED has sought to buffer reliability concerns from the more turbulent developments and negotiations over its program's content and alternatives. Initially, an "assurances group" was set up to provide (as its name suggests) the stability, trust, and commitment stakeholders would need in order to, among other things, accept ecosystem management and the adaptive management of the other CALFED Program components. Assurances that major stakeholders will get the water they need when they need it, however, are nothing less than seeking to establish beforehand the very reliability that the improved management is suppose to achieve; namely, everyone learns how to get the water they need, when they need it. Ultimately, the assurances group disbanded, in part because it was impossible to decouple reliability concerns from negotiations, bargaining, and compromise over CALFED Program elements. To make the same point from the other side, one of CALFED's most noteworthy adaptive management projects would remove power generation dams along a river to be restored for salmon (e.g., U.S. Department of the Interior 2000). But the dams would be removed only in so far as to have a negligible impact on the overall power system reliability.

In sum, the problems associated with adaptive management and high reliability management are remarkably similar in the three ecosystem management initiatives. While noting differences, the senior planner responsible for the Everglades Comprehensive Plan says,

> [I] saw the same kind of tension in CALFED as you find in South Florida. I saw an hour long video from CALFED, a lot of talking heads, and I was struck with the similarity. I'd close my eyes, and every time I heard "Bay–Delta," I'd hear "South Florida." [We] have the same urban, agricultural, and environmental divides. One difference between South Florida and CALFED is that in South Florida the different uses overlap geographically, while in CALFED urban interests are separated geographically in Southern California.

Certainly the same kinds of polarized conflicts are observed across the initiatives: tensions between irrigators/agriculture, power, and environment in the northwest Columbia River Basin; and a highly visible polarization between agriculture and the environment and between agriculture and urban needs in the Everglades, says a South Florida ecologist.

Partial Responses

What have been the responses to the problems with adaptive management in a high reliability context? Are they sufficient to overcome the paradox? Unfortunately, the responses do not save adaptive management as currently defined. Fortunately, some point to a form of management that could work in situations we later term "zones of conflict." The volume and variety of specific responses are impressive and demonstrate an ongoing search for new and better ways to blend or balance the demands of ecosystem management—be it restoration, rehabilitation, or other form of mitigation—with a management dedicated to the highly reliable delivery of water and power supplies. The many different ideas, recommendations, and requests culled from interviews alone show a real hodgepodge of responses, listed in table 3.2. The list can easily be extended by proposals raised in the literature and touched upon in chapter 2 (better pricing, for example).

How can we make sense of this stream of proposals and give each its rightful prominence? We leave the fuller discussion of more important proposals to chapters 5 and 6, after we have developed our own framework in chapter 4 for identifying ecosystem management requirements. Below, we group and introduce many of the proposals in terms of the problems for which they have largely been a response.

Increased Human Population Densities and Resource Extraction

Not surprisingly, if the perceived problem is too many people and too much extraction, the predictable response is fewer of one and less of the other. "How do we restore a major portion of the Columbia River with the

Table 3.2 Miscellany of Partial Responses

- Interagency cooperation and coordination
- Multidisciplinary teamwork
- Restoration and rehabilitation projects
- Watershed planning and management
- Whole-system perspective in planning and management
- Environmental Water Account
- Better modeling and links between different models
- Bringing ecologists and environmental performance measures into the control rooms of high reliability organizations for better real-time management
- Integrated, multi-agency, multi-objective and multispecies planning
- More science and management-relevant research and less politics
- Better monitoring and assessments
- Gaming, simulations, and interagency modeling exercises
- Cross-agency training placements, programs and assignments
- Third-party dispute resolution mechanisms
- New storage and selective water infrastructure
- Water markets
- Managing service demand by adopting other water efficiency and conservation measures
- Blueprints for integrating ecosystem management with high reliability operations
- Flexibility in the Endangered Species Act
- Mechanisms to scale up research findings and scale down whole-system missions and visions
- Stakeholder involvement and public participatory processes
- New institutions and allocation procedures
- Targeted technological innovation
- Less population growth and human consumption
- Management information systems
- Use of environmentally sensitive operational schedules by water and power service providers

population doubling over the next 100 years?" asks the senior Columbia River Basin biologist planner. "[More] storage is a temporary Band-Aid," says a CALFED biologist, "which will only postpone the same problems [we now have] due to rapid population growth." That said, not all ecologists agree on how to respond to population growth. Others believe that while complexity has increased with such growth, some important issues have become clearer and a whole-systems perspective is still possible and useful.

Ecologists differ over the appropriate response to population growth. The lead ecologist at SFWMD raised a criticism of the Everglades Comprehensive Plan's proposal to provide water infrastructure for 12 million people by 2050 in an area where the population is now about half that. "Wouldn't this actually stimulate the growth and thus be counterproductive?" He gave several justifications for the planning target:

> One, the present system was originally designed for 1.5 million people and they thought that was high, and now we have 5–6 million. The underestimation of growth is what got us into this bind. Two, [the

plan seeks] to prevent water wars in the future. Florida has no way to control growth, so the water wars will return if we don't increase supply. We know from experience that in water wars the environment never wins, so there is a justification of more [water].

Although many understandably prefer or wish it otherwise, population growth and intensified resource extraction do spur institutional and technological innovation. While in no way a counsel for complacency, Light et al. (1995) argue that environmental crises can and do foster new ideas and ways to proceed ahead more creatively—but not always or without cost. As an ecologist knowledgeable about the Everglades puts it, "Population and demand increase pressure for people to come up with clever ideas on how to create better adaptive capacity to deal with growing [population] numbers and dependencies." If few stakeholders benefit from the ESA—that is, if its enforcement leaves urban, agriculture, timber, environmental, and recreation users in more uncertainty than before—surely this prods further institutional and technological reform.

While there can be no guarantee that the needed innovations will emerge at the right time, induced innovations and responses have come forth. The irony in ecosystem management is that, while many complain that the costs of the Endangered Species Act exceed its benefits, those costs have had their own benefits. With ESA-engendered conflicts increasing, the 1990s saw, for example, the advent of the PIT (Passive Integrated Transponder) tag, which could be inserted into the body cavities of listed fish, allowing them to be tracked and monitored. The Columbia River Basin biologist planner concludes: "This improvement in measurement has revolutionized the [fish survival] issue. Our monitoring ability and survival estimates have improved by orders of magnitude. The new research has been used for spill decisions, flow decisions, et cetera. ... One surprising result was that [the fish] needed more water in the summer to keep the water temperature down. The power people obviously like that, because that better fits [BPA] schedules." As for institutional innovation engendered by the ESA, we need look no further than the practice and extension of ecosystem management or adaptive management and the drive to find new ways to do both better.

While rapid population growth and extraction have increased the complexity and interconnectedness of the systems, it has also made some matters clearer on the adaptive and high reliability management fronts, at least for some interviewees. Obviously preservation is not enough, says an ecologist with the Nature Conservancy: "This means that we have to do restoration. Which means that all the restoration stuff has to be integrated. And this in turn means all the restoration has to be integrated with [the reliability] operations ... you can't do restoration without flood control, as [when] dechannelizing rivers to their natural floodplains. This is an example of how restoration and operations are intimately tied together."

Science may become clearer as well (Myers 1995). According to Grumbine (1997, p. 43), "regardless of the data gaps—and there are many—usually enough information is available to begin to resolve regional and local [ecosystem management] issues if managers can get access to it." While fish biology is not widely understood in large aquatic ecosystems, some parameters are known, for example, the reproductive life-cycle of some fish. Such parameters, some commentators argue, should be the focus of management (Wilson et al. 1994). "Everyone agrees," says an Everglades ecologist, that better aquatic ecosystem management "has something to do with water flows." The case was put forward by the senior biologist planner in the Columbia River Basin that "The problem is not the science, it's the politics. The science is here, notwithstanding its shortcomings. ... There will be surprises along the way, especially regarding specific technicalities [of restoration or recovery, but] the surprise will not be regarding the functions, the biological parameters."

According to ecologist Donald Ludwig (1996, p. 17), it is "not an environmental problem, nor a lack of knowledge, but a political problem." Others disagree. "It's depressing to hear people involved [in ecosystem management] projects say [the projects] aren't working, and the reason being 'politics,'" argues another ecologist. That's not a good enough reason in her view. The convoluted interconnectedness of all the systems involved—ecological, economic, agricultural, social, and more—first need to be better understood. "Why are we doing this?" asks the Columbia River Basin biologist planner, "We're not spending billions just on getting past the ESA. We're not doing this to delist a few species. ... What is all this about? Let's think about it as a system, not just about the spring chinook where 90% of the attention is."

As the systems have become more tightly interconnected and complex, calls for a "whole-system perspective" in ecosystem management have intensified. "The goal is no longer how many fish are in the river," says the biologist planner, "Now it is producing a system that is able to withstand natural fluctuations. So when El Niño comes along, the fish don't go extinct. Yes, it could be that we have fewer fish at a certain time, but we are really making sure that the populations are robust enough to withstand such fluctuations."

A CALFED biologist working at the Department of Water Resources told of another departmental biologist whose "whole-system view" enabled him to provide real-time management advice on flood control issues related to native fish species, where "the whole-system view" in question integrated both ecological and state water project elements. The "blueprint" jointly proposed by the Nature Conservancy and California's Metropolitan Water District for integrating restoration and water supply operations priorities is called "How to Restore the System," where "system" is deliberately intended to capture both the aquatic ecosystem and the State's high reliability water infrastructure.

Developing a whole-system view leads to a rethinking of system boundaries, a point detailed in later chapters. For example, according to SFWMD's lead ecologist, "the goals of the Everglades Comprehensive Plan depend on the agricultural areas adjacent to the Park remaining there." Why? Because without agriculture, he says, urbanization might sprawl in, posing an even greater threat to the system. Here an altogether larger "pasture" is redefined as a strategic buffer to an altogether different "invasion."

While a whole-system view is difficult to achieve because of its large-scale, long-term nature, its merit is in making more explicit the choice of what decision makers take as their template for restoration, rehabilitation, or other management intervention in the absence of an unrealizable presettlement template. With the drive to clarify goals, objectives, and priorities comes the realization that decision makers have to choose and define these in ways other than appeals to pristine space and time. What decision makers want to restore is "a range of variation instead of the historical template" (the EIS team leader); "what you want to do [in ecosystem management] is increase complexity ... You bring back key species you value" (an ecologist familiar with ecosystem management outside the United States); the template to which the decision maker wants to restore is the normative river, containing all the desirable attributes one wants in a river (the Columbia River Basin biologist planner); decision makers are looking for "desirable changes in keystone species" to take place (SFWMD lead ecologist) or "if critical species are recovering" (a modeler); you want to build in resilience by trying to monitor and improve key variables like nutrients in sediments (an Everglades ecologist). Or, what you do *not* want to do is ignore prevailing climate, disturbance regimes, and vegetation patterns (ICBEMP senior scientist).

The whole-system view, where actually implemented, is thus an important response to the problems of increased human population densities and resource extraction. But as the case evidence makes clear, it has been a partial response at best, largely because it is not at all clear how it integrates the whole-system view or, better yet, integrates this view with the other responses. Chapter 6 provides one such integration.

Availability of Adequate Models and Their Alternatives

Problems with models and modeling have been many, but the response has consistently been to develop and improve individual models and modeling exercises as well as their next best alternatives, assessments, and clear goal definitions in the three initiatives under study. The ecosystem management models and modeling exercises go by different names and acronyms: gaming and Bay–Delta Modeling Forum in CALFED; CRBSUM in ICBEMP; and NSM, AEA workshops, power-modeling and ATLSS in the Everglades. But each has contributed to ecosystem management in their respective areas (for other model-supported efforts, see Gunderson et al. 1995). The discussion of the CALFED gaming exercise is deferred to chapter 5. Those interested in a brief review should turn to the Appendix.

The Bay–Delta Modeling Forum, which meets annually, sponsored the modeling exercise that identified a better way to handle crises like that over the delta smelt.[7] CRBSUM (Columbia River Basin Succession Model), one of many models developed for the Interior Columbia Basin Ecosystem Management Project, has its origin in earlier fire succession models and assumes that multiple succession pathways for plant communities will eventually converge on a stable ("climax") plant community in any given area unless disturbed by natural or human interventions (Keane et al. 1996). It models different types of vegetation throughout the Columbia River Basin under different management and projects how the various disturbances (e.g., fire, pests) would change the succession pathways for each managed vegetation type.

The Everglades predrainage template—the natural system model or its variants—has played an important role in ecosystem management initiatives for that region. For example, a series of workshops were held during the early 1990s, which involved modelers and advocates of both adaptive and high reliability management. During a two-and-half year period, in the words of Gunderson (1999a),

> a dozen workshops were held, involving about 50 technical professionals, mostly biologists and hydrologists. These workshops transformed the understanding of management for the system, a vision that persists to date. At the heart of developing that shared vision was a controversial computer model ... the Everglades AEA [Adaptive Environmental Assessment] model was developed to simulate spatial and temporal dynamics of key ecosystem components. Submodels were developed for hydrologic dynamics, and a set of ecological interactions. Interactions among the hydrology and vegetation, aquatic organisms (fish and invertebrates), alligators, and wading birds were all modeled. The hydrology submodel [a reduced version of the natural system model] became sufficiently credible because of its ability to recreate historical patterns and its application to a subregion, a water conservation area. The ecological submodels were not credible, because the ecological processes that occur on a finer spatial and temporal scale could not be readily aggregated to the scale of the hydrologic model.
>
> The lessons from failures of the ecological submodels led to development of new models, based on aggregating individual dynamics [particularly the ATLSS discussed below]. However, the credibility and generalizability of the hydrology model led to its use in screening policies to identify a subset of policies that deserved a more searching evaluation in terms of feasibility and effectiveness, using other models and other analyses.

The workshop format proved useful later in what were called the power-modeling weekends or "tweak week," during which alternative management scenarios for the wider Comprehensive Everglades Restoration Plan were modeled and evaluated by participants, who included ecologists and water supply representatives.

ATLSS (Across Trophic Level System Simulation) is an ecosystem–landscape modeling approach that uses geographic information system (GIS) vegetation data and hydrology models to drive habitat changes for selected species in south Florida (deAngelis et al. 1998). Results from the hydrologic models are fed into the models of habitat and prey availability for the species, whose behavior patterns are simulated by spatially explicit, individual-based computer models of the species concerned. ATLSS simulations have been used in various ecosystem management alternatives for the Everglades, to compare how proposed water management interventions might affect the spatial patterns of species.

Many more models and modeling exercises have been developed as part of ecosystem management initiatives, and the topic deserves a more thorough appraisal, if not its own book. However, those involved in the more prominent modeling efforts would likely be the first to declare that the efforts have had their limitations; even the success stories are inherently different from those associated with the far more developed and tested engineering models. The latter, while of varying causal accuracy, have proved useful over time in simulating and predicting impacts of hydrology and flow interventions. Ecological modeling and models are typically praised for reasons other than their predictive utility. The insights and better questions that come out of some modeling were mentioned earlier. Yet the efforts have been beneficial for another reason as well.

As we have seen, in the absence of the presettlement template as an automatic guide for ecosystem management, clear and defined goals and objectives for restoration and rehabilitation become the template of choice. Here is where models and their next best alternative, baseline assessments, have been instrumental in advancing ecosystem management, because models and assessments support and drive many of the alternatives and scenarios that follow from the vision, mission, and aims of the ecosystem management initiatives. The scenarios involve predictions based on models and assessments (indeed some scenarios are the only real "test" of the models). More important, scenarios are often justified because they enable the development, comparison, and evaluation not only of different ways to achieve the same goal, but also of different goals and objectives from which to choose in guiding ecosystem management from the outset. Just as scenarios become one way of clarifying models and vice versa, so too do models, modeling, and scenarios clarify goals and compare their possible (hopefully probable) consequences. They thereby help decision makers to decide which goals and which consequences are more desirable. The actual costs and consequences of any subsequent intervention are thus reduced beforehand, making scenario planning the kind of learning-by-doing commonly associated with adaptive management. ICBEMP (United States Department of Agriculture, Forest Service, and United States Department of the Interior, Bureau of Land Management, p. 33) consolidates these considerations for the Columbia River Basin:

> Scenario planning helps describe possible futures, as opposed to a desired future. This process has three main functions: it provides a mechanism

for understanding the integration of management options, it allows people to evaluate the merits, pitfalls, and trade-offs of ecosystem management choices, and it shapes broad perceptions. It is not limited to the resource maximization or cost minimization approaches typical of some multi-resource planning projects. There are two general approaches to scenario planning. The first approach considers varying mixes of inputs and determines the resulting outcomes. This approach provides a mechanism for understanding trade-offs involved in achieving different management goals. Considering the outcomes of several different management scenarios helps managers and stakeholders define what might be possible. Scenario planning in the assessment process uses this approach. The second approach begins with a desired outcome and evaluates different "scenarios" (alternate routes) to reaching the desired outcome. This approach is useful in the decision-making process where differing approaches can be explored to achieve a desired goal or objective. *In a sense, scenario planning is desktop adaptive management that allows managers to experiment without incurring real impacts* [our italics].

Use of models and modeling in comparing and appraising alternative futures, strategies, and options was also found to be pivotal in the Everglades and CALFED ecosystem management. The models used for the Everglades demonstrated, in the words of a lead ecologist, that restoration alternatives helped not only the ecosystem, but water reliability as well. The CALFED gaming exercise, which evaluated different water allocation scenarios and their implications, was able to develop and explore the Environmental Water Account as an option for generating more water flexibility in meeting the state's environmental needs within the context of its other high reliability water requirements.

The three ecosystem management initiatives have had a strong assessment component, largely because of the problems associated with ecological models in their respective cases. There is always good reason to call for more and different assessments under the aegis of improving decision making. Continuous "adaptive" assessment has become an integral part of the Everglades initiative, while ICBEMP has assessment prominently in its very name, as does CALFED's CMARP (Comprehensive Monitoring, Assessment and Research Program). Formal risk assessments of various ecosystem interventions, in particular, are increasingly requested by those seeking more management flexibility than that prevailing under the ESA. More informal models and modeling are also relied upon throughout ecosystem management as it is actually practiced, ranging from the use of spreadsheets to rules of thumb and more tacit forms of knowledge arising out of long experience and familiarity with the ecosystem or species being managed (for examples, see the Appendix). Restoration and recovery do work in some cases, particularly when detailed experience and familiarity with the habitats and species concerned compensate for the lack of formal models and complete assessments. "In [some] instances," *New York Times* science writer William Stevens (2000) reminds us, "nature tends to spring back once the cause of disturbance is removed. Depleted fish populations recover once fishing stops. Forests and grasslands retake abandoned farmland, especially given a helping hand by restorationists. Some ecosystems return to health when cleansing processes like fire, on which they depend, are restored. Rivers tend to cleanse themselves once pollutants are no longer dumped in."

Increasingly Competing Organizational Demands

The response to organizational fragmentation and turf battles has been uninterrupted calls for more interagency cooperation, better coordinative institutions, more effective stakeholder involvement, and third-party resolution mechanisms. A wealth of literature argues for more and better consensus-building in the environmental arena (Bardach 1998) and each ecosystem management case in this book had successful instances of such cooperation. Yet our interviewees were not sanguine about future prospects or past track record. "As a general observation, interagency cooperation or collaboration is an ideal that provides compelling arguments for its existence," says the ICBEMP senior scientist, "but [it] is fraught with many potential failings." His colleague, the EIS project leader says, "My experience is that interagency work has been some of my best times and also some of the worst." She adds, "But if agencies can accept each other's different role, then you can get further. Land managers can get further with their work when others consider their involvement in that. If managers want to make it work they will; if they don't, it won't."

Whatever the merits of interagency cooperation, it has yet to change the status quo, at least in the way that the ESA has. Accordingly, even with cooperation, the pressure for more substantive institutional and technological innovation remains an ever-present necessity. Interviewees noted that many of the regulatory agencies have, under pressure, become more multi-species in orientation, including NMFS and the Fish and Wildlife Service. Various state resource agencies or departments of environmental quality with broader or more flexible mandates have been created to fill the vacuum left by state fish and game departments focused on species and fixed management strategies. Ecologists we interviewed call for new institutions for ecosystem management, such as "meshing institutions" that provide regulatory and planning agencies with new research findings, science, ideas, and input for ecosystem management. "In cases of successful adaptive assessment and management, an informal network seems always to emerge," writes Gunderson (1999b):

> That network of participants places emphasis on political independence, out of the fray of regulation and implementation, places where formal networks and many planning processes fail. The informal, out of the fray, shadow groups seem to be where new ideas arise and flourish.

It is these "skunkworks" who explore flexible opportunities for resolving resource issues, devise alternative designs and tests of policy, and create ways to foster social learning. How to develop and foster shadow networks is a challenge for most inwardly looking North American land management agencies.

As for stakeholders, the public consultation process led, in the view of the ICBEMP senior scientist, to a better product:

> They had really good input and in some cases comparable to peers [e.g., in peer review]. That was because people were better informed than we initially supposed, they were far better read in the literature, and they understand issues far better. They draw inferences themselves rather than be told [what to think]. They can link things together, and they can take data and challenge you. All of this forces you to ground your own case better. They found errors before peers could ...

Barry Johnson, of the U.S. Geological Survey, sums up the stakeholder involvement:

> Although adaptive management does not require consensus on objectives before implementing management experiments, a lack of well-defined objectives that reflect stakeholder values seems to result in less support for the process. Future applications of adaptive management might be improved by including more open discussion of differences in stakeholder values with the goal of developing some objectives, perhaps very broad, that most stakeholders can agree to. In addition, managers need to find ways to incorporate the nonscientific knowledge and data that stakeholders possess into the adaptive management process. (Johnson 1999, p. 3, online)

A notable stakeholder group actually wanting to resolve the paradox has been the least commented upon in the literature, but is crucially important for this book: high reliability organizations. Habitat and ecosystem restoration (including rehabilitation and mitigation) have become part and parcel of the mandates of the high reliability organizations (e.g., SFWMD, California's Department of Water Resources, and the Bonneville Power Administration). In fact, we were told that BPA was the world's largest fish and wildlife agency, with a budget more than that of NMFS and FWS combined.[8] Nevertheless, the extent to which ecosystem and reliability are conceptually integrated, let alone optimized together, remains a major concern. As the ecologist described the joint blueprint proposal of the Nature Conservancy and the Metropolitan Water District,

> it is evident that there is a need for a wider framework than CALFED for integrating flood control and restoration activities. In addition there are all these restoration and recovery projects that could run outside of CALFED or even counter to [it]. So what are you going to do when there is no steelhead recovery plan in CALFED? But there will be one,

so how will it integrate? The operators, both agriculture and the water districts, feel cheated by CALFED. They thought it was the big umbrella, but it turns out it isn't. The idea of the blueprint is to recouple all recovery and restoration to ERP [Ecosystem Restoration Process] and make ERP the 'rebuttable assumption' for the regulatory agencies. It's the hurdle the regulatory agencies have to jump over when they issue their recovery programs.

The drive is strong in ecosystem management to find the equivalent of the budget ceiling that no one dare exceed rather than the budger base from which everyone negotiates upward. Yet in the absence of "binding" interagency cooperation, super-agency control or blueprint, the pressure has been to seek the next best alternatives, such as management structures that resolve policy and management differences earlier on or, when that is not possible, effective third-party resolution mechanisms. Interviewees were always pleased when a management unit lower in the hierarchy was able to make a decision that did not require resolution at a higher level. Where interagency conflicts needed to be resolved, means for third-party conflict resolution were sought but not always available.

Rising Demand for Multiple Services from Ecosystems

The primary response to the zero-sum nature of competing demands for services, whether related to restoration or reliability, has been to propose new infrastructure to increase the supply of water or find flexibility in the existing water and power system. Ecosystem management in the Everglades exemplifies the first strategy, while the second is largely the thrust of initiatives in the Columbia River Basin and CALFED. In fact, the latter two initiatives are conspicuous in their lack of new infrastructure, with BPA talking about decommissioning dams in the face of rising power demand and CALFED deliberately delaying a decision on new storage facilities until after the first seven years of program implementation. While more storage is often touted as something that will make everyone happy, including environmentalists (seeking to settle the water wars), chapter 5 discusses the major problems with storage, many of which are by and large unacknowledged.

Since major new water supplies are not on the horizon in most cases, the search is for more flexibility in the existing system so as to meet competing service demands better. There is a pot-pourri of different strategies: more integrated planning, programming, and implementation, all continuously updated and redesigned in light of new information (e.g., CALFED, Everglades Comprehensive Plan, ICBEMP); dedicating more water for ecosystem restoration and the environment (e.g., CALFED's Environmental Water Account); better management of the multiple differences in perspectives, objectives, priorities, and evidence (e.g., third-party dispute resolution and the TNC/MWD blueprint); developing tradeoffs and synergies within a whole system or strategic perspective (e.g., the gaming and power-modeling exercises); and enhancing operational flexibility for restoration or reliability goals (e.g., new technologies and better modeling). Chapters 5 and 6 return to a fuller discussion of the more important proposals.

An alternative to meeting the demand for services through either new infrastructure or better use of the existing facilities and systems is to redefine the services to be supplied through ecosystem management or high reliability management. The team leader for the Columbia River Basin EIS argues, for example, that the forests should no longer be seen as sources of traditional extraction and services, but rather as sources of more sustainable goods and products: "Whether it is timber you want to harvest or high quality water because you are taking care of the riparian and aquatic systems, you'll get more sustainable services." One way to redefine services is to insist that all ecosystem management activities, and the services they generate, be multipurpose. That is, any single resource should have multiple uses. The same water in CALFED's Environmental Water Account could be used and reused for different purposes during the course of a season or year, including assisting the recovery of a species now and as urban drinking water later elsewhere. Indeed, the EWA itself, and not just its water, is a single resource having multiple purposes. As we will see, environmental stakeholders like the idea of an Environmental Water Account because it accepts in principle that the environment should have its own checkbook. Agricultural stakeholders like the idea because they would finally know how much was in that checkbook (plus the EWA could create incentives for environmentalists to support new water storage facilities, if simply for environmental purposes). Urban stakeholders like the idea because some of the checks would be for environmental restoration and improvements with direct and positive impacts on urban water quality. We take up these notions of redefinition and resources with multiple uses in the next chapter, which sets out a comprehensive framework for not only adaptive and high reliability management regimes, but for the other forms of ecosystem management as well.

Conclusion

Such have been the problems and responses associated with adaptive management and ecosystem management in a working environment that places a very premium on the reliable provision of ecosystem services. None of the responses on their own or in tandem is enough. The gap between what is available and what is needed is best reflected in the calls made by our interviewees and in the literature for more leadership, full-cost pricing and internalizing externalities, institutional overhaul, better politics, more public education, and, always, calls for more trust or, at least, less distrust and polarization (e.g., Grumbine 1997; Cortner et al. 1998, p. 162; Slocombe 1998, table 3; Lee 1999; Schindler and Cheek 1999). These are good things to seek, but it all sounds a bit like Napoleon who when asked what he wanted in his generals replied, "Luck." "The simple part is over," concluded the senior Everglades planner, "the complexity of the effort has expanded geometrically." "Do we have the courage to do something dramatic or just continue nibbling around the edges?" asks the Columbia River Basin biologist planner. We turn now to our answer.

4 Recasting the Paradox through a Framework of Ecosystem Management Regimes

We now provide a parsimonious framework for recasting the paradox so that it can be acted on. Our framework of ecosystem management regimes is used in the following chapters to resolve the impasse between ecologists and engineers. In so doing, it integrates engineering more positively into ecosystem management than is currently done. The goal of ecosystem management is a twofold recoupling: where decision makers are managing for reliable ecosystem services, they are also improving the associated ecological functions; and where they are managing for improved ecological functions, they are better ensuring the reliability of ecosystem services associated with those functions. In practice, improvements in ecosystem functions may range from preservation or restoration of self-sustaining processes to the rehabilitation of functions by reintroducing to the ecosystem something like the complexity and unpredictability they once had.

The recoupling of functions and services that have been improved varies by the type of management (more formally, the management regime) relied on by decision makers, where the principal task facing the decision maker is to best match the management regime to the ecosystem in question. A "regime" can be thought of as a distinct and coherent way of perceiving, learning, and behaving in terms of variables discussed more fully below and summarized in table 4.3 at the end of this chapter (for more on policy and ecological regimes in ecosystem management, see Norton 1995, p. 134; Berry et al. 1998; for a discussion of regime theory, see Kratochwil and Ruggie 1986).

To summarize our argument, while ecosystems are internally dynamic and complex, they also vary along a gradient in terms of their human population densities, extraction, and other significant features discussed in chapter 3, such as differing models, competing organization, and multiple-use demands. In response to changes along the gradient, ecosystem management passes through thresholds (the most important being limits to learning) as decision makers move from one management regime to another. The thresholds, in fact, are best thought of as gradual transitions between modes and models of learning about ecosystems. While being precipitated or punctuated by sudden flips or discontinuities in the biophysical realm due to population and extraction pressures, the thresholds in ecosystem management signal that one way of learning to manage and managing to learn is superseded by a very different way of doing so. Since ecosystems are almost always located in landscapes that are heterogeneous, not just ecologically, but politically, socially, and culturally as well, the decision maker faces a decision space where both the ecosystems and learning about them are uncertain, complex, and incomplete.

Under these conditions, how do decision makers restore patterns of complexity and unpredictability to species, habitat, and ecosystem functions and at the same time ensure the steady, safe, and reliable supply of services from that ecosystem? How can ecologists and engineers work better together to save the environment and service reliability?[1] The first step of our answer is to better match the ecosystem to an appropriate management regime. To justify such an approach, we must start by establishing just how saturated ecosystem management is both with political, social, and cultural values and with considerations of scale.

Political, Social, and Cultural Values

Chapter 2's review underscores the importance ecologists and others place on political, social, and cultural factors in ecosystem management. Ecologists are most concerned with connecting ecosystem functions and services in improved biophysical ways. But the goal itself and its how, when, and where are quintessentially social, political, and cultural in nature. Engineers increasingly understand the importance of such value choices. What we take to be the key management regimes, how the regimes fold in stakeholders, how they define ecosystem services (even functions) through the different management regimes, and our very choice of templates to guide the recoupling of services and functions, all and more are rooted in political, social, and cultural choices.

Difficult decisions for each management regime are clearly value-laden throughout. For self-sustaining management, the choice is to preserve self-sustaining functions by fitting services to those functions so that the former follow from and do not undermine the latter, that is, to use a popular metaphor, services are the "interest" on nature's capital (see Folke et al. 1996; Grumbine 1997, p. 42). For adaptive management, the choice is to restore functions in ways that mimic as best as possible the presettlement template, historical processes, or a related template, but for a more differentiated set of services. For high reliability management by ecologists and engineers (again broadly writ) it is to isolate and control, if not redefine, the ecosystem functions necessary to maintain an altogether wider and more problematic set of services. And for what we call case-by-case management, it is to capitalize on ecosystem improvements and service reliability so they enhance each other for specific places and times. Nothing in this chapter suggests that these management regimes determine in some value-neutral way in which management and ecosystems could be best matched to meet predefined needs. Values are core to that match and definition.

Scale and the Decision Space

Most experts writing on ecosystem management stress the importance of scale. We have already discussed the positive features of large-scale adaptive management experiments and the negative features of large-scale engineering interventions. Other examples of the importance of scale are legion. Lackey (1998, p. 24) argues that "a set of decisions to maximize benefits in managing a thousand-ha watershed within the Columbia River Watershed may well be very different than decisions for the same small watershed that were designed to maximize the benefits over the entire Columbia River Watershed." While many such arguments warn against committing the familiar fallacies of composition,[2] people do need continual reminding of their importance. Patterns and processes that occur at the landscape level may not be visible at the ecosystem level or below, and the dynamics of the landscape cannot be understood by adding up the properties and processes of the ecosystems. Scaling up and scaling down of management-relevant information thus is almost always difficult (Root and Schneider 1995).

Scale is important in two ways for the framework of ecosystem management regimes. First, the framework is ecosystem-based, and a management intervention in one part of the ecosystem can have interaction effects on other parts of the same ecosystem. Second, the interaction effects of the management intervention can extend beyond the ecosystem to another ecosystem, nearby or farther afield. For instance, protecting a national park and its ecosystems may attract increasing human population densities surrounding the park (thereby generating new service reliability needs and opportunities). Thus, the two kinds of scale effects of interest here are (i) the geographical scale of the ecosystem and its landscape and (ii) the scale of the intervention in an ecosystem or landscape. The same intervention could have a large or small effect depending on the size and composition of the ecosystem. Similarly, the impact on an ecosystem of an intervention could be large or small, depending on the scale of the intervention. The decision space in which the decision maker operates is to a large extent complex, uncertain, and incomplete precisely because of these combined geographical and intervention effects.

Table 3.1 in chapter 2 underscores the overriding influence of scale. Decision makers find it hard to assess whether their intervention has worked, due to the lag time between making an error through an intervention, and the discovery of that error and the magnitude of the error once discovered. As ecologists stress, on many occasions interventions, intended or not, have rapid and observable negative impacts (Carpenter et al. 1995, p. 325). Other times the impacts become visible only much later, often too late to remedy (e.g., Vitousek et al. 1997, p. 498). Probe this dilemma further and the effect of scale becomes clear in defining just why the decision maker's space for intervening is so fraught with difficult and paradoxical choices.

It is scale that determines whether or not, or the degree to which, errors are reversible or irreversible, lethal or non-lethal, linear or nonlinear, direct or indirect, continuous or discontinuous, and high (low) probability or high (low) consequences (see, e.g., Ayres 1995; Kaufmann and Cleveland 1995; Mintzer 1995; Myers 1995; O'Neill et al. 1996). In fact, these are the dimensions of the interaction effects of most concern to the decision maker. The scale of the hazard, in terms of both its magnitude and its probability, is a function of the scale of the landscape and the scale of the intervention leading to the hazard. In case it needs saying, the why, what, and how of considering low-probability, high-consequence, irreversible, lethal effects to be important in ecosystem management is a decidedly political, social, and cultural choice.

Since these interaction effects are usually uncertain, complex, and incomplete for many real-world management horizons, the decision maker's decision space is especially dynamic, both ex ante and post facto. "When 'cumulative effects' are investigated, the interactions of even a few stressors produce horrendously complicated and counterintuitive effects on ecosystems" (Schindler 1996, p. 18). Small effects may be irreversible and lethal; large effects may be indirect and non-linear; others may be cascading and direct; many more may simply be unknown in terms of their probability and consequence, whether high or low. In this way, the principal feature of the ecosystem management decision space, at least at the landscape and larger scale, is its potential for time-sensitive but unknown negative externalities working against achieving the twofold management goal. Decision makers know they have to do something now, but will it do any good? Could it actually cause harm?

In such a decision space, it is tempting to equate urgency with certainty. As discussed in the preceding chapter, some ecologists, engineers, and others believe they know enough to take action before things get worse. After all, as one ecologist we interviewed put it, "if you don't do anything, it is the fastest road to extinction." When there are only 50 adult panthers left in the Everglades, the decision maker knows a problem is at hand.[3] But what should the decision maker be doing instead? That's the truly difficult question to answer. Indeed, many ecologists argue strenuously that we do not know enough to manage ecosystems effectively. Not only are the science, engineering, or both lacking in this view, but the political, social, and cultural factors giving rise to multiple and contested definitions of ecosystems and ecosystem health (or its cognates) also complicate decisions as to what to do now and in the days ahead. Furthermore, the factors are often themselves scale-sensitive, as when Berry et al. (1998) described the decision space in the following terms: "no one is in control," "there is no clear picture of public expectations or understandings ... No one knows about society's long-term wants and needs," and "Large-scale ecosystems have the likely prospect of involving many more interests, jurisdictions, authorities and so forth, most of which one will have to engage and overcome to be successful." Lackey (1998, p. 22) summarizes the matter:

> Ecosystem management problems have several general characteristics: (1) fundamental public and private values and priorities are in dispute, resulting in partially or wholly mutually exclusive decision alternatives; (2) there is substantial and intense political pressure to make rapid and significant changes in public policy; (3) public and private stakes are high, with substantial costs and substantial risk of adverse effects (some also irreversible ecologically) to some groups regardless of which option is selected; (4) the technical facts, ecological and sociological, are highly uncertain; (5) the "ecosystem" and "policy problems" are meshed in a large framework such that policy decisions will have effects outside the scope of the problem. Solving these kinds of problems in a democracy has been likened to asking a pack of four hungry wolves and a sheep to apply democratic principles to deciding what to eat for lunch.

It is important to stress, however, that while the decision space is full of uncertainties, complexities, and unfinished business, not everything in that space is uncertain, complex, and incomplete. Indeed, some risks are obvious in the sense that their hazards and probabilities are fairly known, at least for some areas of the world. Biodiversity is under threat. We know that for a fact. We know that conventional irrigation destroys land unless active remedy measures are taken. Old-growth forests are disappearing, and by and large we know why. In fact, such certainties are the main engine driving decision makers to undertake ecosystem management on a larger scale and with greater urgency, in order to more effectively address these other problems. It is, however, precisely that urgency and scale, both in terms of addressing large landscapes with large-scale interventions and in fostering the human values to support such interventions, that make the decision maker's space irremediably dynamic. What to do then? How can we manage better if the decision space in which we operate is so dynamic? Our answer is to propose a framework for ecosystem management that accepts this dynamism, but in the process differentiates the decision space for more useful purposes later.

Typology and Implications of the Decision Space

The intersection of values and scale generally, and of interaction effects, coupling, and decision making specifically, are summarized in a more formal representation of the relevant decision space. The rest of this chapter uses that representation, adapting a typology developed first by the sociologist, Charles Perrow (1984). For our purposes, the decision maker's space for management interventions is composed of different systems—ecological, social, political, cultural—that have two chief dimensions: coupling and interaction. The former is a structural feature of the decision space. The latter is a feature of action or activity in that space. A tightly coupled decision space is (1) highly time-dependent, allowing for few delays or unexpected contingencies; (2) fairly invariant in terms of the sequence of activities or functions required (i.e., B depends on A having occurred first); (3) inflexible in the way its objectives are by and large achieved (not only is the sequence of activities restricted, but there is only one way to achieve the overall goal desired); and (4) characterized by little slack and resources available to bridge delays, stoppages, and the unexpected when they do occur. In a loosely coupled decision space, delays are not only possible, but common; sequences of activities are by no means invariant (e.g., in most degree programs it does not matter much when certain requirements are met before getting the degree); there are many ways to achieve a common goal; and sufficient slack exists to tolerate a measure of waste without imperiling the decision space in the process.

Both tightly and loosely coupled decision spaces can be, in turn, dynamically or linearly interactive. Dynamically interactive spaces are those with uncertain, complex, and incomplete sequences of activities, functions, or services whose consequences (reversible/irreversible, lethal/nonlethal, direct/indirect, continuous/discontinuous, and low/high probability) are invisible, incomprehensible or incomplete in significant respects. They are uncertain because causal processes are unclear or not easily understood. They are complex because the sequences and activities are numerous, varied, and interrelated. They are incomplete because the sequences and activities are, more often than not, interrupted and thus left unfinished in important respects. Sequences and activities in a linearly interactive decision space are, by contrast, more familiar and expected and are quite visible, comprehensible, or perceived to be complete, even if unplanned or unintended. The dimensions of coupling and interaction produce a typology of four types of systems for decision makers. Examples from the organizational literature for the four cells are dams (tight-linear), most outsourced manufacturing (loose-linear), nuclear power plants (tight-dynamic), and universities (loose-dynamic).

In ecosystem terms, the four cells define the type of systems and associated decision spaces in which the ecosystem decision maker operates (table 4.1). All four decision spaces are found in the ecological literature (Peterson et al. 1998). Ecologists, engineers, and other decision makers, however, increasingly find themselves in a tightly coupled, dynamically interactive decision space where they are facing unknown but potentially massive effects of decisions. This decision space stands in sharp contrast to the conventional thinking about natural resource management that was often in tightly

Table 4.1 Examples of Decision Spaces

		Interaction	
		Linear	Dynamic
Coupling	Tight	for example, brittle systems without redundancies, where causal pathways are largely comprehensible and predictable	for example, brittle systems without redundancies, where causal pathways are not largely comprehensible and predictable
	Loose	for example, resilient systems with redundancies, where causal pathways are largely comprehensible and predictable	for example, resilient systems with redundancies, where causal pathways are not largely comprehensible and predictable

coupled, linear terms (that is, manage forest production and offtake on a planned schedule) or the wish to treat actual natural resource systems as if elements in the system were loosely coupled, but linear in effect (that is, decision makers could experiment in the forest without the ecosystem collapsing).

Ecosystem management is quite clearly a response to conventional thinking. "Ecosystem-based management is not a simple, linear activity," concludes Slocombe (1998, p. 485), "It is based on large areas that are diverse ecologically, economically, and socially, and complexly connected and interacting." As preceding chapters discussed, population growth and extraction have made the human and physical landscapes over which ecosystem management takes place so tightly coupled and interactively complex that the Ecological Society of America takes the complex interconnectedness and dynamic character as fundamental elements of ecosystem management itself. "What has happened [in response] is that managers want that one-size-fits-all scenario, which they find so appealing," said one of our scientist interviewees, adding, "It basically treats all land as having the same risks." Yet no amount of pull that decision makers feel for clear and defined objectives (presuming a tightly coupled, linear interactive system) or push they feel when confronted with the prospect of persistently complex, uncertain, and interrupted choices, makes their decision space any less dynamically interactive in nature. Here, the decision maker never knows if the risk-averse strategy being pursued is conservative enough or, for that matter, too conservative (e.g., Wilson et al. 1994, p. 299). In this decision space, the decision maker may never really know if the objective of species recovery has actually been achieved, unless a political or social judgment has been made that this is so.

In contrast to the "blueprint" sponsored by the Nature Conservancy and Metropolitan Water District (chapter 3), there is no blueprint for coupling improved ecosystem functions and reliably provided ecosystem services in a tightly coupled, dynamically interactive decision space. Instead, the need is to identify management options that better enable decision makers to differentiate their decision space in ways that accept rather than ignore its dynamic nature. The most effective way to do this is by matching different types of ecosystems with their most suitable management regimes. While the threat of, for example, irreversible effects of interventions is present throughout the landscape and its ecosystems, questions of irreversibility are more important for certain management regimes than for others.

This book avoids proposals that assume the best way to make choices in a dynamic decision space is to translate them into linear terms. A typical linear proposal is the *deus ex machina* to save us from the paradox. For example, Berry et al. propose the "creation of a single organization to manage and coordinate ecosystem management research in the Pacific Northwest" (1998, p. 75). They do so largely because in their view the current state of ecosystem management is so polarized and fragmented across state and federal lines, agencies, and jurisdictions that the "system is out of control" (Berry et al. 1998, 69) in the Columbia River Basin. Such a proposal for a region-wide management board looks as if it takes scale and values seriously. But does it really? If ecologists agree on one thing with respect to scale, it is that all scales are important because any one scale contains dynamics that may be unobservable from the scales above and below it (chapter 2). If all scales are important, there cannot be "the" appropriate scale for management, as Berry et al. and indeed others seem to assume. The motivation for their proposal and others like it appears to be the perceived need to get direct (i.e., tightly coupled and linear) control over a decision space that is everything except tightly coupled and linear in its scale and value heterogeneity.

Accepting heterogeneity has a second important consequence for this book. The scale and value considerations just described govern our choice of looking at ecosystems from a landscape perspective. We wanted a scale of analysis and management that contains whole ecosystems, introduces unavoidable ecological and nonecological diversity into the management decisions, and can be used in an abstract conceptual sense as much as a real land-use category when it comes to talk about connecting services and functions in our twofold management goal. The landscape is one such commonly recommended category and is widely accepted as an appropriate level for framing ecosystem management options (Pickett and Cadenasso 1995, p. 332; Ayensu et al. 1999, p. 686; Rhoads et al. 1999, p. 302). Accordingly, our discussion of the framework begins at the landscape level.

The Framework

Start with a hypothetical landscape and its ecosystems. Ecological definitions of "landscape" and "ecosystem" differ (chapter 2), but it is sufficient here to define a landscape as a heterogeneous land area with multiple ecosystems. An ecosystem in that landscape is a local biological community and its diverse patterns of interaction with the landscape and the wider environment. Our landscape's ecosystems are varied, though their exact nature (freshwater, marine, mountain) is less important than that they differ along the interrelated dimensions found in chapter 3 to be crucial in setting the limits of adaptive management to ensure reliable ecosystem services:

1. Human population densities in the ecosystems range from low to high per unit of land.
2. Resource extraction from the landscape's ecosystems ranges from virtually nonexistent to high (as measured in, e.g., land values, barter terms of trade, and pollution discharges).
3. Availability of adequate models to explain and predict relationships important for management purposes ranges from relatively few, if any, to many such models.
4. The mix of ecosystem and organizational health considerations varies across the ecosystems, ranging from those in which ecosystem health predominates in management to those where organizational health (i.e., competing organizational demands often for high reliability services) predominate.
5. Ecosystems vary from those representing a source of multiple, diverse functions with few services to humans (e.g., the ecosystem could be managed for different consumptive and nonconsumptive services, but is not) to those ecosystems whose resources each provide multiple consumptive and nonconsumptive services (e.g., water is used for recreational, agricultural, environmental, and urban purposes).

These five dimensions define any landscape's gradient of ecosystems, their management regimes and the thresholds between one management regime and another. Simplified, the framework is summarized in figure 4.1.

Figure 4.1 positions ecosystems at one point in time, the present. For heuristic purposes, ecosystems in the landscape have been grouped into four "states" or categories of control and use: those relatively undominated by people and their needs; those colonized to some extent by humans; those that have moved from colonization to increasingly competitive extractive uses and human domination; and lastly, ecosystems where human domination and regular extractive use for high reliability purposes are their eminent features. The real world, of course, has mixed cases, most prominent being that some ecosystems are relatively unpopulated but nonetheless heavily extracted due to increasing demand for services from outside the ecosystem. Such mixed cases tell us that the ecosystem's placement in one of the four categories can never be completely tied to population numbers or to extractive uses alone (more below).[4]

Some examples of the four ecosystem categories are (in right to left order in figure 4.1) wilderness or remote areas; large government parks; areas where population growth, natural resource use, and demand for environmental services are in conflict; and the full-blown urban ecosystems of cities and towns. One would be mistaken, however, in equating the categories with

Self-Sustaining Management	Adaptive Management	Case-by-Case Management	High Reliability Management
e.g., "wilderness areas"	e.g., national parks, consumptive use of ecosystem services such as recreation	e.g., zones of conflict where population, resources, and the environment increasingly compete	e.g., urban ecosystems, pastoralist ecosystems

Human colonization of an ecosystem — Human domination of an ecosystem — Human control of multiple ecosystems for high reliability

Ecosystems least populated by humans ⟷ Ecosystems most populated by humans

No extraction of resources ⟷ High extraction of resources

Few causally adequate models ⟷ Many causally adequate models

Ecosystem health, with organizational implications ⟷ Organizational health, with ecosystem implications

Multiple sources/few services ⟷ Single resource/many services

Figure 4.1 Framework of ecosystem management regimes.

wilderness areas, parks, zones of conflict, and cities, if we were to conclude that the thresholds that divide ecosystem categories are largely legal and administrative in nature. The reason why the examples differ is due to increased controls and uses to which humans put the ecosystems in question. These latter factors depend on political, social, and cultural considerations. It is these factors that provide the context for the thresholds and the implications of these thresholds for managing the ecosystems (more about thresholds later).

Five consequences of our twofold management goal frame this chapter's discussion on how the management regimes and ecosystem categories are to be matched for more effective ecosystem management. These points are introduced here because of their importance, after which we proceed to discuss the management regimes specifically.

The Ever-Present Requirement

First, all ecosystems in our landscape are in need of some management recoupling. Why? Because services are present everywhere in our landscape, and there is no place where services and functions are not decoupled and in need of recoupling, once improved. As one moves to the right of the gradient, improved ecosystem functions and services becoming increasingly decoupled: rivers that once had floodplains now have channels; arid and semi-arid ecosystems become irrigated farmland; and wetlands turn into runways. Even those few ecosystems that would be classified as best mimicking a presettlement template or other normative cognate are in need of management, if "simply" to establish and then maintain the infrastructure needed to preserve the areas as unique places in the landscape's natural heritage (which is the service these functions perform for us). Here management seeks to reduce harmful services or, better yet, to avoid those harmful services entirely. What is decoupled and in want of effective recoupling in these otherwise self-sustaining ecosystems is the obligation of decision makers to intervene to preserve the ecosystem, else the disconnection between services and functions worsen under pressures of population growth, extraction, and demands for more and different services (i.e., as the ecosystem moves from one category to another along the gradient).

Fatal Flaw

Second, from the perspective of our framework, the fundamental policy question in ecosystem management is how to manage. And the fundamental policy implication is that the exclusive use or recommendation of any one management regime across all categories of ecosystems within a heterogeneous landscape that is variably populated and extractively used is not only inappropriate, it is fatal to meeting the twofold goal of effective ecosystem management. Preference of one management regime over all others all the time and everywhere is the worst kind of linear thinking in a persistently dynamically interactive decision space. Our framework presupposes that decision makers and other stakeholders accept the provisional and contingent nature of the decision space in which they operate and seek the optimal management regime for the ecosystems they are to manage. Sadly, there will always be those who believe the decision space really is a linear world of obvious problems with obvious solutions.

In practical terms, this means that the environment's trinity discussed in chapter 2—getting right the prices, institutions, and incentives as a precondition for effective ecosystem management—has causality backwards. The decision maker must first match the optimal management regime to the ecosystem being managed and then draw out implications for the kinds of institutions, incentives, and pricing mechanisms that need to be in place to undertake that management. As will be seen below, the institutional arrangements for undertaking adaptive management or high reliability management are not only different; they are orthogonal in important respects unless specifically designed otherwise when resolving the paradox.

Fortunately, when it comes to information needs, the right match between management regime and ecosystem enables us to identify indicators of effective ecosystem management. As detailed in chapter 3, identification of useful indicators for monitoring depends crucially on having a causal theory or theories about the ecosystem and its management regimes. That theory for each management regime is described below. That there are different management regimes implies that the indicators of meeting the twofold management goal will be determined by the regime itself, the relevant ecosystem category, and the overall location of both in terms of the fivefold gradient. Indicators at or around the thresholds will, however, be ambiguous for reasons that will become clear.

Paradise Lost and Regained

Third, the nature and mix of ecosystem services and functions vary along the gradient, since the challenge of meeting the twofold management goal varies by management regime. In this way, the twofold management goal of recoupling improved functions to reliable services is just that, a goal. How it is realized in practice depends on the management regime and the way in which that regime has (re)defined the services and functions to be recoupled.

Services end up being redefined all along the gradient under increasing pressures of population, extraction, and the other dimensions. For example, an ecosystem that started on the gradient's left as an incredibly rich source of irreplaceable biodiversity ends up on the gradient's right, disaggregated into extractable resources, each having its own next best substitute and each of which has many different uses (O'Neill et al. 1996, p. 23; Luke 1997). The whole ecosystem is decoupled into discrete "natural resources." Ecosystem functions shift from a dynamically interactive whole with few "natural" services into specific functions linked to increasingly specific services. Functions move from many to few, if any, self-sustaining ones, while services move from natural to highly disaggregated ones. Ecosystem management, thus, moves from the presettlement template of preserving and maintaining self-sustaining ecosystems to goals, objectives, and scenarios to rehabilitate functions associated with service reliability. Thus engineers play an increasingly important role when we move from left to right. In practice, a goal of ecosystem management, to ensure both self-sustaining functions and reliable services, gets translated as one moves to the right of the gradient, to improving functions without compromising services. Figure 4.2 sets out a stylized version of these changes over the four ecosystem categories and across the gradient.

We follow chapter 2 in defining restoration and rehabilitation as the main ways of improving ecosystem functions. What is different here is that the framework makes much more explicit the role of engineers in restoration and rehabilitation when it comes to case-by-case management and high reliability management. As we saw in the preceding chapters, next to restoring self-sustaining processes, improving ecosystem functions typically means (re)introducing to ecological functions a degree of complexity than they may have had historically, or need in order to maintain keystone species and processes. More complexity might also be needed to make the ecosystem in question more resilient in the face of the unexpected. In some cases, restoring or rehabilitating ecosystem functions by introducing complexity may even increase the reliability of ecosystem functioning, which has important implications for the other half of the twofold management goal, the service reliability associated with these functions. Not all improvements can be managed the same everywhere, however. Restoration, as we shall see, is

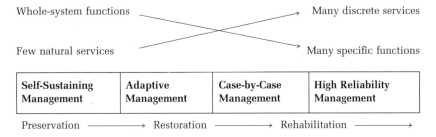

Figure 4.2 Service and function redefinition (from left to right on the gradient).

more suitable for self-sustaining and adaptive management, while rehabilitation is more consonant with case-by-case and high reliability management regimes.

How the twofold management goal is defined along the gradient is not the only thing that changes; so too do the functions and services to be managed. What does it mean to say that functions and services are "redefined" in response to changes along the gradient? Redefinition can come from redefining the boundaries of the systems being managed under pressure, to see them from a whole-system perspective. Pastures and forests on their own are characterized one way, but once they are treated as part of a pasture–buffer–forest system, the function of hydrologic and species transfers between the subsystems are added to the analysis and management equation. The ecosystem functions associated with a biodiversity "hotspot" on its own would be managed one way, but if this hotspot were to be managed as part of a wider, surrounding system, or what Daily (1999) calls countryside biogeography, the functions would be managed another way. New opportunities for different management strategies and options arise in such remapping of boundaries. The pressure to redefine boundaries from subsystem to whole system arise in large part precisely because the factors operating along the fivefold gradient—increased population and extraction, better models, the demand for more services, and increased organizational pressures—all work to highlight interdependencies between subsystems where few were perceived before.

Services too are being redefined along the gradient. National forests that once were conceived largely in terms of providing timber are now being reconceived in terms of providing sustainable services, according to the EIS team leader for the Integrated Columbia Basin Ecosystem Management Project. "Originally set aside for other reasons, national parks and wilderness areas are," as Callicott et al. (1999, p. 32) summarize, "being redefined as biodiversity reservoirs, the primary function of which is to provide living room for interacting and mutually dependent nonhuman species populations." It hardly needs to be said that many of these populations are under threat as the landscape moves to the right of the gradient.

Services and functions are sometimes being redefined together. Many environmentalists believe ecosystems stop where the city begins. Inside the city limits nature is destroyed; outside the city nature is imperiled—imperiled, moreover, by swelling urbanization. An article on restoration ecology in *Science* (Dobson et al. 1997) plotted the relation between the spatial scale of natural and human-made disasters (in square kilometers) and their estimated recovery time (in years). The figure shows urbanization to be a disaster worse than oil spills, floods, modern agriculture, or the blast of an atomic bomb, but not as bad as a massive meteor strike. On the other hand, in a more recent *Science* article, Kloor (1999) argues that urban ecosystems are just that, ecosystems with their own rich and diverse set of species and processes. Some ecologists now see a positive relation between cities and ecosystem functions, even seeing in exotic species some positive functions and services. (Similar findings have been found of biodiversity on heavily used lands, be they communal areas in Zimbabwe or pastures in the urbanizing Green Heart of the Netherlands; see also Daily 1999.)

In other cases, services and functions are not so much redefined as expanded. One program element of the Everglades Comprehensive Plan is to store freshwater "bubbles" in the saltwater aquifers. The function of the aquifer would remain storage, but the services derived from it would be increased. Similarly, proposals to restore a forest to its more historical pattern of low-intensity, high-frequency fires may entail incorporating tree species not found there originally. In other instances, redefinition is simply a matter of time. As is often pointed out, today's native species were once invader or exotic species. "We are focusing a lot of attention on invasions of species within very small scales of time, but Earth's intermingling of organisms from different continents and seas has been occurring millions of years" (Young 1999, p. 901; see also the extended discussion over natives and exotics in Callicott et al. 1999).

A good way to begin understanding how services and functions are redefined over time and space is to think of the ways in which ecosystem services actually have a positive, rather than negative effect on functions. Replacement of sensitive species by more tolerant ones in the same functional group may improve the functioning of the ecosystem. Selective timber cutting may result in an increase in overall ecosystem productivity (Richardson 1994, figure 1). Moreover, in human-dominated ecosystems it can be difficult to separate the service from the function, or the management of the ecosystem from how it is actually used. Placing the village's footpaths alongside irrigation canals is one way in which villagers manage those canals, that is, people walk by regularly and are thus likely to see problems before they get out of control. Using the levee road atop the Delta island is one way in which managers keep track of what is going on, or not going on, both on the land and on the watersides of the road (chapter 5). Residents surrounding a federal or state forest may take firewood from the forest or serve as an early warning system for fires, thereby enhancing forest health in the process (Fortmann 1990). In such cases, how the resource is used is how it is managed, and the service safeguards the function(s) from which it is derived.

How do such changes in functions and services fit into the four management regimes and across the gradient? Here is where redefining the management goal combines with redefining the services and functions to be managed. Start at the left of the gradient. Relatively few pressures of population, extraction, and other factors serve to undermine the presettlement template of self-sustaining functions with their "natural" services as the guide for ecosystem management. The decision maker moves to an adaptive management regime as soon as pressures introduce uncertainty and complexity over which are the appropriate interventions for recoupling services and functions, that is, as soon as the presettlement template ceases to be an unproblematic guide to management because more and different people, extraction, and services are now involved. Not only have more functions and services become decoupled, but the services and functions in question are now more differentiated. The ecosystem itself has changed. The decision maker faces more demands with respect to what needs to be learned before intervening to meet the twofold management goal. Also, because population, extraction, and other pressures on the ecosystem have increased, the decision maker has fewer options to reduce or eliminate services, which is a preferred alternative in self-sustaining ecosystems.

One main reason why redefinition of services and functions along the gradient leads to different management regimes is that the nature of risks and the demand to manage them change along the gradient as well. If risk is the magnitude of a hazard multiplied by its probability, clearly the nature of hazards changes as well as their probabilities as one moves rightward on the gradient. In self-sustaining management there are hazards (e.g., the forest ecosystem can burn away in a lightning fire) and some notion of probability (e.g., the historical frequency of lightning fires in such a forest), but there is little reason to "manage" the forest directly if the ecosystem is treated as a self-contained ecosystem meriting hands-off management. By the time one gets to high reliability management, the hazards are very different (e.g., keeping the water from the forest aquifer clean for use by people). Probabilities, too, are much more specific, where the focus of high reliability management is to directly manage both the magnitude of the hazard and its probability of occurrence. Redrawing the boundaries as a way of redefining services and functions is important, then, for another reason: namely, such redefinitions, in addition to creating new opportunities for management, also involve changing the set of hazards and probabilities that they treat as risks to be managed. People manage not only because services and functions have been redefined, but also because the latter entails new kinds of hazards, probabilities, and opportunities that require different management strategies (e.g., from a whole-system perspective).

The most dramatic way that ecosystem functions and services are redefined along the gradient and by management regime is what happens at the high reliability management side of the gradient. As discussed in chapter 3, high reliability management of huge water supply systems has undeniably harmed ecosystem health. If we were to leave it at that, high reliability management would surely remain the enemy of ecosystem health just as self-sustaining management is tautologically the best for ecosystem health.[5] But our framework is intended to be a prescriptive as well as descriptive guide to improved ecosystem management. We argue more fully below that high reliability management has a positive role to play—and necessarily so—in meeting the twofold management goal. Such a role must be so, if only because the right side of the gradient, with its ecosystems of high population densities, extraction, and other demands, makes it inconceivable that any management regime there could avoid the demand for reliability in ecosystem services. There high reliability management will have to be part of the solution, notwithstanding its clearly problematic track record. For instance, fire and fuel-load management are increasingly seen as active parts of the high reliability management of government forests. So too are integrated pest management practices an important part of achieving reliable production from agro-ecosystems (Matson et al. 1997, p. 508). Equally important, even on the left of the gradient, we see the demand for high reliability management in the form of establishing and managing the very stewardship infrastructure needed to preserve self-sustaining ecosystems reliably over time (on the need for stewardship, see Grumbine 1994, p. 34; on the need for advanced technology in this stewardship, see Harte 1996, p. 28). With stewardship infrastructure introduced, services and functions are redefined for self-sustaining ecosystems as well. We return to this point more fully later in the chapter.

Cornered

A fourth implication of our framework's twofold management goal concerns thresholds. The thresholds that distinguish the management regimes have continuous and discontinuous features when it comes to meeting the goal. Much of the literature on ecosystem management underscores the importance of thresholds in ecosystem functions and services (Grumbine 1994, p. 32; Common 1995, p. 103; Kaufmann and Cleveland 1995, p. 110; Max-Neef 1995; Myers 1995; Opschoor 1995, p. 138; Page 1995, pp. 146–147; Walker 1995; Daily et al. 1996; El Serafy and Goodland 1996). Many thresholds are considered to be ecological responses to stressors that change or, indeed, flip the ecosystem, or parts of it, from one state to another (Arrow et al. 1995). Moreover, the stressors of interest are by and large those of the gradient, for example, population numbers, extractive uses, demand for ecosystem systems, and competing organizational considerations. Such thresholds vary empirically from landscape to landscape and ecosystem to ecosystem. Furthermore, in a decision space of certain, causally comprehensible, and complete information about ecosystem services and functions, each landscape's thresholds would be defined just as empirically in terms of the stressors and processes for any given ecosystem state and transition.

Our hypothetical landscape, however, always faces a potentially tightly coupled, dynamically interactive decision space. It is one where information about ecological thresholds is rarely certain, complete, or straightforward at least for time-sensitive decision-making purposes. In our framework, the relevant thresholds are less in the ecosystems themselves than in different management regimes. Here the other dimension of the gradient is crucial: the models underlying the management regimes. Each management regime is most usefully thought of as a theory of knowledge generation and its limits (Roe 1997). In this view, it is best to think of the thresholds as limits to learning and the transitions to new ways of learning. As population growth, extraction, and other factors change, the decision maker eventually has to confront the limitations of the management approach relied on in terms of learning to manage better and managing to learn better when it comes to achieving the twofold goal.

Examples of the limits to management learning abound. They are found in all manner of asymmetrical arguments made by ecologists and others about managing the environment under pressure along the gradient. We are told it is easier to mismanage an ecosystem than it is to manage it better (e.g., Ludwig et al. 1993; Ludwig 1996). Humans "have been about as effective in managing ecosystems as plankton have been in managing lakes," as one ecologist put it (Schindler 1996, p. 18). In this view, ecosystem collapse is more certain than ecosystem sustainability. Negative externalities are more predictable than positive ones. Nature on its own is too complex to control. But our mismanagement of nature has unleashed forces we no longer can control. We must manage the planet's resources better, but we cannot expect technology to help us in doing so. Economic growth is never a sufficient condition for improving the environment, while economic growth's potential irreversible impacts on the environment are always a sufficient condition for adopting the precautionary principle (e.g., Arrow et al. 1995). So much is uncertain that anything is possible, and thus everything must be at risk. Whatever humans touch they make worse, is Barry Commoner's Third Law of Ecology (Light et al. 1995, p. 137). When we asked an ecologist what native species have benefited from ecosystem changes induced along the gradient, he was at a loss for words.[6]

The precautionary principle is the clearest illustration of a limit to learning in ecosystem management. The principle, which calls for decision makers to make no intervention unless they can prove it does no harm, looks utterly reasonable in a world where high reliability management is destroying the environment and where most adaptive management has yet to deliver results. On the other hand, acceptance of the principle as a guide to management places a palpable limitation on the way we learn about doing no harm, since it is not possible to prove that an undesirable outcome will never happen. It is logically impossible to prove a negative (Duvick 1999). Thus, managers have to work with a principle which, if carried to its logical conclusion, means undertaking few, if any, new management interventions involving risk. Our framework assumes that managers will reach a point—literally a threshold—where they will begin looking for new ways of learning rather than remain at such a limit or in such asymmetrical argumentation. This is the threshold at which they move from what they can and do manage to what they want to manage but nonetheless is unmanageable under prevailing management thinking (Roe 1998). What was reasonable under one regime is no longer sensible under another regime.[7]

Sometimes the management-based limits to learning are less obvious. In their well-regarded "Economic Growth, Carrying Capacity, and the Environment," Kenneth Arrow and his colleagues (1995) argue

> A single number for human carrying capacity would be meaningless because the consequences of both human innovation and biological evolution are inherently unknowable. ... A more useful index of environmental sustainability is ecosystem resilience. ... Even though ecological resilience is difficult to measure and even though it varies from system to system and from one kind of disturbance to another, it may be possible to identify indicators and early-warning signals of environmental stress. ... The problem involved in devising environmental policies is to ensure that resilience is maintained, even though the limits on the nature and scale of economic activities thus required are necessarily uncertain.

In other words, what is being recommended is an altogether uncertain measure, resilience, of inherently uncertain ecosystems, on which to base environmental policies whose nature and consequences are themselves necessarily uncertain. Such a "challenge" certainly puts decision makers at the limits of what they can expect from the kind of experiment-based adaptive management recommended by Arrow et al. (1995).

Learning-related management thresholds are, of course, related to within-ecosystem biophysical thresholds. In fact, notions such as carrying capacity and ecosystem resilience are themselves a tangled amalgam of biophysical and learning-based limits, where agro-ecosystem innovation and technology can improve both but not in guaranteed or sustainable ways. Ecological thresholds are often taken into account through the learning process associated with the specific management regime being relied upon. Discontinuities in learning can be introduced or catalyzed because of the dynamics of the ecosystem under pressure to change along the gradient (on discontinuities, see Folke et al. 1996, p. 1019). Annual grasses disappear, and herders realize they cannot continue to graze as they did before; fish are no longer upstream, and business as usual is no more. Obviously, sudden transitions or flips have sudden and profound impacts on achieving the twofold management goal in terms of the operating management regime.

It would, however, be a mistake to think of thresholds *solely* in discontinuous terms either as limits to management learning or as sudden ecosystem changes. "The danger of emphasizing discontinuous change is reminiscent of the parable of the frog that just sits still and dies in water which is heated

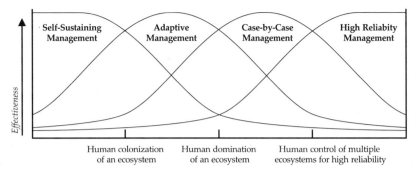

Figure 4.3 Management regime effectiveness curves.

slowly, but will attempt to jump out if placed into a boiling pot," counsel Kaufman and Cleveland (1995, p. 110). So too for our management regimes. The reality is that while one management regime is optimal for a given ecosystem category, all four management regimes are potential alternatives for ecosystem management anywhere in the landscape. Decision makers, given their ambitious twofold goal and limited budgets, are open to all management strategies. This in no way argues that all management regimes are equally optimal for a specific ecosystem or that any one management regime is always optimal for all ecosystems. Rather than flipping from one management regime to another as in a paradigm shift along the gradient, decision makers will give stronger emphasis to the management regime that better suits the circumstances under which they now need to learn and manage (for a discussion of paradigms and paradigm shifts in ecosystem management, see Grumbine 1994, p. 35; Lackey 1998). Thus, figure 4.1 is not the whole picture. Figure 4.3 plots stylized effectiveness curves for the four management regimes, showing both the optimal management regime for each ecosystem category as well as the three nonoptimal alternatives present for meeting the twofold management goal (the thresholds appear as the intersection of adjacent optimal effectiveness curves). For example, the left tail of the effectiveness curve for high reliability management could be in the form of the stewardship infrastructure needed to preserve the self-sustaining ecosystem. Similarly, the right tail of the self-sustaining management curve could take the form of the restored patch habitat associated with high reliability waterworks. We return to such examples below. More formally, however, while there are many different ecosystems, the major management regimes for learning about them are far fewer in the framework.

Out of Bounds

The twofold management goal has a fifth important implication. While the framework operates within a context of scale and interaction effects at the inter-ecosystem and intra-landscape levels, transboundary effects important for achieving the twofold management goal can extend well beyond the landscape. Negative externalities produced elsewhere can and often do impact ecosystem functions and services. These transboundary effects take different forms, ranging from global climate change to the destruction of habitat of migratory birds. Stephen Farber (1995, p. 105) writes, "It is global and transboundary ecosystem impacts that are the most worrisome from a management perspective." Our framework does not guide the assessment of how to better manage these wider, outside-landscape transboundary effects, since they take us beyond issues of ecosystem management as commonly understood (for an analysis of managing these global environmental issues in light of local governance systems, see Ostrom et al. 1999). That said, the landscape level remains a potent unit of analysis for large-scale externalities, most prominent being global climate change, as many experts now agree that global climate change models need to be "regionalized" if suitable policy and management interventions are to be developed (e.g., Ayensu et al. 1999).

Four Management Regimes

With these five implications of the twofold management goal in hand, we now turn to a detailed discussion of the framework's four management regimes. Nothing in what follows implies that the four management approaches are the only theories available for resource management (for others, see Roe 1998; Miller 1999). Those presented are the prevailing theoretical perspectives recommended in current professional literature for the major ecosystem categories identified, and reflect the political, social, and cultural values associated with Western-style ecosystem management. Thus, our focus is on how adaptive management, high reliability, complex adaptive systems, and related concepts operate within ecosystem management. This may well be different from how the concepts are defined and used in other fields, or even by other ecologists and engineers. Each of our theoretical perspectives can be clearly defined in terms of the five dimensions of the gradient, and each posits a management approach which (it contends) increases the chances of successfully meeting the twofold management goal for the ecosystem category in question. Two of the management regimes, adaptive management and high reliability, were introduced in preceding chapters and are summarized more formally below. We touched on self-sustaining management in the form of the presettlement template, but this management regime is also described more fully below. As case-by-case management has not yet been described, but given its importance for subsequent chapters, more space is devoted to its explication. In the process, it will become clear why the dimensions of modeling, health, and (re)sources health (in figure 4.1), along with human population densities and extraction, set the management regimes apart from each other, constitute the criteria for choosing these theories over others, and explain the nature of the thresholds between management regimes. Indeed, only the four management regimes capture fully the kinds of changes under way along all five dimensions. For ease of exposition, we start at the ends of the gradient.

High Reliability Management

Those in the control rooms of large-scale power- and water-generation systems are preoccupied with ensuring a service reliability that comes from the characteristics of high reliability management (table 4.2), but how can high reliability come from, let alone be imposed over a tightly coupled, dynamically interactive decision space in which decision makers seek to better couple ecosystem functions and services? How can decision makers suppose that the very organizations that have harmed the environment can actually be instrumental in achieving our twofold management goal? The answers require a theory.

High reliability theory was developed by organization theorists interested in how complex organizations and institutions maintain their activities in situations where failure, error, and accidents are highly probable (Rochlin 1996). The primary question is (Demchak 1996; La Porte 1996), How do some institutions, with complex systems and in predictably unstable environments, still manage to continually meet peakload production in a reliable, safe fashion? In addition to the water and power control systems of chapter 3, high reliability organizations (HROs) studied include air traffic control systems, nuclear power plants, electricity companies, hospital intensive care units, and naval air carriers. Many heavily populated ecosystems, it bears repeating, require high reliability management as well, if only to ensure a steady stream of ecosystem services, including but not limited to water supply and quality.

Rochlin (1993) summarized the principal features of high reliability management in his "Defining 'High Reliability' Organizations in Practice: A Taxonomic Prologue." The work of other high reliability theorists, particularly

Table 4.2 Principal Features of High Reliability Organizations

- High technical competence
- High performance and oversight
- Constant search for improvement
- Often hazard-driven flexibility to ensure safety
- Often highly complex activities
- High pressures, incentives and shared expectations for reliability
- Culture of reliability
- Reliability is not fungible
- Limitations on trial and error learning
- Flexible authority patterns
- Positive redundancy

La Porte (1993, 1996), is used to supplement and extend the list to 11 related features:[8]

1. *High Technical Competence.* High reliability organizations are characterized by the management of technologies that are increasingly complex and which require specialized knowledge and management skills in order to safely meet the organization's peakload production requirements (Rochlin 1993, p. 14; La Porte 1993, p. 1). What this means in practice is that the organizations are continuously training their personnel, with constant attention devoted to recruitment, training, and performance incentives for realizing the high technical competence required (Roberts 1988, figure 3; La Porte 1996, p. 63). To do so means not only that there must be an extensive database in the organization on the technical processes and state of the system being managed, but that this "database" includes experience with differing operating scales and different phases of operation, the proposition being that the more experience with various operating scales and the more experience with starting and stopping risky phases of those operations, the greater the chance that the organization can act in a reliable fashion, other things being equal (La Porte 1993, p. 3; Perrow 1994, p. 218).

2. *High Performance and Oversight.* Technical competence in an HRO must be matched by continual high performance. The potential public consequences of operational error are so great that the organization's continued success, let alone survival, depends on reliably maintaining high performance levels through constant, often formal oversight by external bodies. As Rochlin (1993, p. 14) puts it, "The public consequences of technical error in operations have the potential for sufficient harm such that continued success (and possibly even continued organizational survival) depends on maintaining a high level of performance reliability and safety through intervention and management (i.e., it cannot be made to inhere in the technology)." Accordingly (Rochlin 1993, p. 14), "Public perception of these consequences imposes on the organizations a degree of formal or informal oversight that might well be characterized as intrusive, if not actually comprehensive." La Porte (1993, p. 7) adds, "Aggressive and knowledgeable formal and informal watchers [are] IMPORTANT. Without which the rest [i.e., high reliability] is difficult to achieve." Note that "oversight" does not mean close supervision of personnel within the HRO; in fact, overly close supervision is inimical to achieving high reliability (Schulman 1993a). Rather the oversight in question typically comes from external bodies that demand high reliability from the HRO's senior managers, who in response allocate resources to achieve that reliability.[9]

3. *Constant Search for Improvement.* A feature related to high technical competence and constant monitoring is the continued drive to better operations in high reliability organizations. Personnel constantly strive to improve their operations and reduce or otherwise avoid the hazards they face, even when (or precisely because) they are already performing at very high levels. "While [high reliability organizations] perform at very high levels, their

personnel are never content, but search continually to improve their operations" (Rochlin 1993, p. 14). They seek improvement "continually ... via systematic gleaning of feedback" (La Porte 1996, p. 64). Notably, the quest is not just to do things better, but to reduce the intrinsic hazards arising from the activities of many HROs (T. R. La Porte pers. comm. 2000).

4. *Often Hazard-Driven Flexibility to Ensure Safety.* Not all HROs face highly consequential hazards. But many do, and these hazards drive them to seek flexibility as a way of ensuring safety. "The activity or service [of these HROs] contains inherent technological hazards in case of error or failure that are manifold, varied, highly consequential, and relatively time-urgent, requiring constant, flexible, technology-intrusive management to provide an acceptable level of safety [i.e., reliability] to operators, other personnel, and/or the public" (Rochlin 1993, p. 15). The more hazardous the operations, the greater the pressure to ensure high reliability of those operations.

5. *Often Highly Complex Activities.* Not unexpectedly, then, the more complex the actual operations and activities performed, that is, the more inherently numerous, differentiated, and interdependent they are, the greater the pressure to operate in a highly reliable fashion (e.g., Rochlin 1993, p. 15). What this means in practice is that high reliability organizations often find it "impossible to separate physical-technical, social-organizational, and social-external aspects; the technology, the organization, and the social setting are woven together inseparably" (Rochlin 1993, p. 16). In such organizations, its technology, social setting, and units are extremely difficult to tease apart conceptually and practically. Such complexity characterizes many activities of many HROs. Note the qualification "many." Not all activities in an HRO are complex nor do all HROs center around complex activities (T. R. La Porte 2000, pers. comm.). As our effectiveness curves imply, high reliability management is found over a range of conditions. The point here is that the more complex (and the more hazardous) the operations of an organization in combination with the other features discussed above and below, the greater the pressure to manage in a highly reliable fashion.

6. *High Pressures, Incentives and Shared Expectations for Reliability.* HRO activities and operations must meet social and political demands for high performance, with safety requirements met in the process, and clear penalties if not (Rochlin 1993, p. 15). One way to do so is to ensure that those who do the management work live close to the system they manage—they fly on the airplanes they build or guide, they live downwind of the chemical plants they run or on the floodplains they manage, and their homes depend on the electricity and water they generate (Perrow 1994, p. 218).

7. *Culture of Reliability.* Since the HRO must maintain high levels of operational reliability, and safely so, if it is to be permitted to continue to carry out its operations and service provision, a culture of reliability comes to characterize the organizations (Roberts 1988, Figure 3; Rochlin 1993, p. 16). This means in practice that the organizations often exhibit clear

discipline dedicated to assuring failure-free, failure-avoiding performance (La Porte 1993, p. 7).[10] Such a culture does not mean the organization is saturated with formal safety regulations and protocols, which as with overly close supervision end up working against the achievement of high reliability. A culture of high reliability is one in which core norms, values, and rewards are all directed to achieving peakload performance, safely, all the time, informally as well as formally (T. R. La Porte 1997, pers. comm.). As Rochlin (1993, p. 21) puts it, "the notion of safe and reliable operation and management must have become so deeply integrated into the culture of the organization that delivery of services and promulgation of safety are held equally as internal goals and objectives: neither can be separated out and 'marginalized' as subordinate to the other, either in normal operations or in emergencies."

8. *Reliability Is Not Fungible.* Because of the extremely high consequences of error or failure, high reliability organizations cannot easily make marginal tradeoffs between increasing their services and the reliability with which those services are provided (Rochlin 1993, p. 16). "Reliability demands are so intense, and failures so potentially unforgiving, that ... [m]anagers are hardly free to reduce investments and arrive at conclusions about the marginal impacts on reliability" (Schulman 1993b, pp. 34–35). There is a point at which the organizations are simply unable to trade reliability for other desired attributes, including money. Money and the like are not interchangeable with reliability; they cannot substitute for it. High reliability is, in brief, not fungible.

9. *Limitations on Trial and Error Learning.* Given the above, it is not surprising that high reliability organizations are very reluctant to allow their primary operations to proceed in a usual trial-and-error fashion for fear that the first error would be the last trial (Rochlin 1993, p. 16). They are characterized by "inability or unwillingness to test the boundaries of reliability (which means that trial-and-error learning modes become secondary and contingent, rather than primary)" (Rochlin 1993, p. 23). While HROs do have search and discovery processes, and elaborate ones, they will not undertake learning and experimentation that expose them to greater hazards than they already face. They learn by managing within limits and, if possible, by setting new limits, rather than testing those limits for errors (T. R. La Porte 1999, pers. comm.). As Rochlin (1993, p. 14) puts it, high reliability organizations "set goals beyond the boundaries of present performance, while seeking actively to avoid testing the boundaries of error." Trial and error learning does occur, but this is done outside primary operations, through advanced modeling, simulations, and in other ways that avoid testing the boundary between system continuance and collapse. Chapter 6 returns to the importance of managing within limits and setting new ones.

10. *Flexible Authority Patterns.* High reliability organizations "structur[e] themselves to quickly move from completely centralized decision making and hierarchy during periods of relative calm to completely decentralized and flat decision structures during 'hot times' " (Mannarelli et al. 1996, p. 84).

In particular, these organizations have a "flexible delegation of authority and structure under stress (particularly in crises and emergency situations)" (Rochlin 1996, p. 56), where "other, more *collegial*, patterns of authority relationships emerge as the tempo of operations increases" (La Porte 1996, p. 64). When this ability to rapidly decentralize authority under stress is combined with the persistent drive to maintain flexibility and high levels of competence in an HRO, the management emphasis is to work in teams based on trust and mutual respect (T. R. La Porte 2000, pers. comm.). In this way, emergencies can be dealt with by the person on the spot, whose judgment is trusted by other members of the team that works together in these and less charged situations. We return to the importance of such teams and trust in chapters 5 and 6.

11. *Positive Redundancy.* Last, but certainly not least, HROs are characterized by multiple ways in which to respond to a given emergency, including having backup resources and fallback strategies. This can happen in three ways:

> First, functional processes are designed so that there are often parallel or overlapping activities that can provide backup in the case of overload or unit breakdown and operation recombination in the face of surprise. Secondly, operators and first-line supervisors are trained for multiple jobs including systematic rotation to assure a wide range of skills and experience redundancy. Thirdly, jobs and work groups are designed in ways that limit the interdependence of incompatible functions. (La Porte 1996, pp. 63–64; see also Rochlin 1993, p. 23)

Again, if we had to physically locate the above features, we would be looking for the control room with line operators and engineers forming the core of the HRO, as is found in our San Francisco Bay–Delta, Columbia River Basin, and Florida Everglades water supply projects. These rooms are the one place in our case studies where we will find the technical competence, complex activities, high performance at peak levels, search for improvements, teamwork, pressures for safety, multiple sources of information and cross-checks, and the best example of the culture of reliability, all working through technologies and systems that build in sophisticated redundancies to buffer against potential failure and catastrophe. In this way, the complex, uncertain, and open-ended decision space of the decision maker is displaced by the who, what, where, and how of control by line operators. To make this work, performance measures have been and are being developed, most recently including ecological performance measures being of use in engineering and maintaining large technical systems such as the water supply system (Schulze 1999). The control room in these systems is the decision space for high reliability management.

Resource management within an ecosystem context based on these features of high reliability can be found in situations defined by the dimensions at the right of figure 4.1's gradient; for example, high population densities or regular, widespread extractive uses. Indeed, high reliability management has been called for in densely populated urban ecosystems (Roe 2000) and is found in extensive-grazing pastoralist ecosystems (Roe et al. 1998). Both require not just a healthy ecosystem, but also healthy institutions and organizations adept at dealing with complex technologies and systems that extend beyond the ecosystem to the landscape level and beyond. As one high reliability theorist, Paul Schulman (1996, p. 74), puts it, "reliability often becomes synonymous with a proxy variable—organizational health." It is not a luxury good that only the rich or wealthy nations can afford or aspire to.

It bears repeating that for ecosystems at this right end of the gradient, high reliability management must be part of better recoupling of reliable services to improved functions. At one level, this seems implausible. If self-sustaining management is tautologically good, then high reliability management must be tautologically bad. The apogee of high reliability management would appear to be the total decoupling of services from functions, with services being defined in such a way as to exclude the very need for self-sustaining ecosystem functions. Again, what started as a holistic ecosystem has now been reduced to decoupled resources associated with a set of services and functions that have been redefined, in large part precisely because the hazards and probabilities they pose can be better managed in a highly reliable fashion.

However, the redefinition of services is less than half the story. High reliability also has the positive role described earlier, for functions are also being redefined along the gradient, and this redefinition has important implications for the positive features of high reliability. Consider how it is not only high reliability engineers and line operators who have failed to recouple improved services and functions, but ecologists as well. As we saw in preceding chapters, a common recommendation by ecologists and others is that the best way to recouple services and functions is by reducing population growth and per capita consumption; that is, move the ecosystem from the right to the left of the gradient. All that would achieve, however, is to treat recoupling as if it were a matter of reducing services or redefining them entirely in terms of functions. Yet these recommendations seek to forever decouple targeted services from functions or targeted services from other services. Since this is not a politically, socially, or culturally acceptable option for most ecosystems at the right of the gradient, such proposals ironically end up mirroring the very decoupled situation already in place. Except by decoupling excess(ive) humans from nature, advocates of reducing population growth and per capita consumption seldom offer positive ways to recouple improved services and functions to meet the twofold management goal wherever along the gradient. Chapters 5 and 6 are intended to fill that gap.

What such proposals miss is that functions, and not just services, are being redefined all along the gradient. But how can the functions of, say, organic matter accumulation or nutrient cycling be redefined? Are they not permanent categories in the ecological sciences? The answer is that the context in which these functions are understood changes along the gradient, which in turn means reinterpreting the role these functions have in meeting the twofold management goal. By the time we reach high reliability management and the ecosystems for which it is most appropriate, we have left the usefulness that the presettlement template and whole-ecosystem perspective have for the other management regimes. In highly reliability management, the goal is the restoration or recovery of more improved functions, where the operative phrases are "more" and "functions"; that is, such interventions can be effective in terms of meeting the twofold management goal even if the entire ecosystem is not restored to what it was when there were no (European) settlers, or 100 years ago, or when "everything was better." Indeed, the ultimate redefinition of functions and services comes about when the instant decision makers understand that even self-sustaining ecosystems are on permanent life-support, requiring a huge stewardship infrastructure of high reliability management to maintain it into perpetuity. To see why, we need a better understanding of self-sustaining management, its tie to the presettlement template, and the ecosystems for which they are most suitable.

Self-sustaining Management

At the end of the gradient farthest from intensely managed ecosystems, with their discrete services and specific functions, are ecosystems with far fewer people, extraction, and services, where the ecosystem's dynamic interactions are very difficult to model, simulate, or tease apart and where whole ecosystem health has priority over the organizational health considerations found in high reliability management. Such ecosystems are, of course, not truly pristine. The planet may yet harbor a few such ice caps, hyperarid deserts, and mountain peaks, but our hypothetical landscape has none of these. Even the landscape's wild or remote areas are in some need of some improvements to better meet the twofold goal of linking self-sustaining functions to reliable services. An isolated area may have been mined in the past or a relatively unmolested viewscape may be threatened by increasing numbers of tourists and urbanites. It is important to note that appropriateness of self-sustaining management is not limited to isolated or "untouched" ecosystems. There are many empirical examples of many "forgotten" ecosystems that have thrived because no one was managing them, such as the border zone between the Koreas and areas omitted from official zoning regulations or government programs (D. E. Rocheleau 1999, pers. comm.; J. Hukkinen 2000, pers. comm.).

In short, functions in parts of the ecosystem may no longer be as self-sustaining as others found elsewhere, while the few services produced by the ecosystem may not be reliably assured. Fortunately, since many ecosystem functions are already self-sustaining, the presettlement (predisturbance, historical) template serves as a fairly useful guide to restoring those functions that fall short of being self-sustaining. It is true that ensuring those services that follow naturally from the ecosystem continue reliably into the future may well require stewardship infrastructure to preserve the ecosystem. Nonetheless, the point here is that at this end of the gradient, decision makers are seeking ecosystems that internally manage themselves and remain self-correcting, preferably without the infrastructure but with it if required (e.g., Willers 1999). For many people, these ecosystems must remain endogenously self-sustaining so as to minimize, if not permanently forestall, human mismanagement of them. Here, then, the ideal is self-sustaining management.

It is important to stress that such ecosystems are by definition healthy ecosystems, precisely because they approximate the presettlement template of a self-organizing ecosystem. "Natural ecosystems are tautologically resilient," Colin Clark puts it (1996, p. 112). The coupling of services and functions is treated as ideal, because services have been redefined to be only those that do not interfere with already self-sustaining functions (e.g., the only "use" of the ecosystem may be its existence value). In this way, critics are right to point out that ecosystem resilience is a fuzzy concept in human-dominated ecosystems (e.g., Ayres 1995), if simply because true resilience is by definition possible only without people. While high reliability management recouples services to functions largely by redefining functions (treating them as disaggregated features of a less-than-whole ecosystem), self-sustaining management recouples largely by redefining services (by treating them as aggregated features inseparable from the whole ecosystem).

The relevant analytic framework for understanding the presettlement template of a resilient, self-organizing ecosystem is complex adaptive systems theory. Such ecosystems are complex adaptive systems because their numerous, varied, and interdependent components—primarily ecological—interact in ways that enable the system to self-organize and improve its chances for survival (its resilience) within an ecosystem or landscape. Only a fraction of the literature on complex adaptive ecosystems treats humans as a force for self-sustaining activities and, for the most part, self-regulation and self-correcting ecosystems work best, in the view of this literature, when human beings are absent or use the area lightly. Unsurprisingly, many ecologists and other natural scientists recommend that ecosystems within the landscape remain as self-sustaining and self-correcting as possible.

Complex adaptive systems theory has been developed to understand principles underlying the dynamics of diverse systems ranging from ecosystems to economies. Such systems are thought to resemble each other in that they are difficult to understand and manage from the outside, and their evolving nature provides a "moving target" that is extremely difficult to model. Work is currently being done to develop fairly simple models of individual behavior, say, of ants, which, when combined with simple rules, give rise to surprisingly complex but coordinated behavior (e.g., Johnson 1999). Full-blown, robust models of complex adaptive systems, however, do not exist. Indeed, the lack of adequate models is the main reason why ecosystem health and integrity are so difficult to define or measure precisely (chapter 2), even when political, social, or cultural consensus exists for the twofold management goal.

The difficulty in modeling comes from the fact that complex adaptive systems, including ecosystems, share (in the view of some commentators) three characteristics: evolution, aggregate behavior, and anticipation. All of these are seen to generate the system's capacity to self-organize its behavior and thereby ensure its own internal self-sustaining management (see Holland 1992). "Evolution" refers to the ability of parts of the ecosystem to adapt and learn. The systems of interest exhibit complex adaptive processes with many parts and widely varying individual criteria for effectiveness. "Aggregate behavior" is the ability of an ecosystem to exhibit behavior that is not simply derived from the action of its parts, such as cycles of flow of materials and energy. "Anticipation" is the ability of the parts to develop rules that anticipate the consequences of certain responses. This attribute makes the emergent behavior of complex adaptive systems both intricate and difficult to understand. In this way, a complex adaptive system builds and uses its own "internal models," a characteristic that defies traditional modeling methods. Anticipation, along with evolution and aggregate behavior, enable new behavioral rules and macro/meso relationships to emerge, often spontaneously, from discrete individual behavior and micro interactions. These rules and relationships, in turn, are core to the ecosystem's resilience. As Arrow et al. (1995, pp. 92–93) put it, such resilience "is a measure of the magnitude of disturbances that can be absorbed before a system centered on one locally stable equilibrium flips to another.... For example, the diversity of organisms or the heterogeneity of ecological functions have been suggested as signals of ecosystem resilience." Such properties are the essence of self-organizing complex adaptive systems.

Adaptive Management

Most comparatively unpopulated grasslands, lakes, rivers, and forests in our landscape have been directly influenced by past human intervention: the forests are second-growth, rivers' water levels are regulated by structures, lakes contain introduced species, and the mix of flora and fauna in the grasslands has changed because of previous livestock utilization. In these ecosystems, the numbers and impact of people have been considerable in comparison to the few wild or remote areas that exist elsewhere in the landscape, though far fewer numbers of people and impacts have been incurred than in the fully human-dominated ecosystems where high reliability management is found.

Adaptive management has its greatest salience and applicability in these human-colonized but not intensely dominated ecosystems. For it is here where the presettlement template ceases to be unproblematic in guiding and developing management interventions to rehabilitate or restore the ecosystem or parts of it back to more self-sustaining functions. Why? The ecosystem has more people, extractive uses, and ecosystem services, which in turn have become more differentiated. The drive to adequately model behavior increases accordingly, and ecosystem health must now to be balanced against organizational considerations in place to extract and provide those services. In short, more services have become decoupled from self-sustaining functions and more functions have ceased to be self-sustaining. The presettlement template or its cognates remains important, but now the choice is to restore or rehabilitate functions for a more varied, redefined set of services than before. Nature gives way to natural resources; natural resources give way to exploitation for more varied and specific services.

What is adaptive management in such situations? Again, there are many varieties of adaptive management and chapter 3 outlines the features most frequently recommended. The business of designing adaptive management strategies involves four basic issues, according to Carl Walters (1986, p. 9), an early promoter of adaptive management:

- bounding of management problems in terms of explicit and hidden objectives, practical constraints on action, and the breadth of factors considered in analysis;
- representation of the existing understanding of managed systems in terms of more explicit models of dynamic behavior that spell out assumptions and predictions clearly enough so that errors can be detected and used for further learning;
- representation of uncertainty and its propagation through time in relation to management actions, using statistical measures and imaginative identification of alternative hypotheses (models) that are consistent with experience but might point toward opportunities for improved productivity;
- design of balanced policies that provide for continuing resource production while simultaneously probing for better understanding and untested opportunities.

From this position, the principal driver of adaptive management from the perspective of its proponents is the dynamic uncertainty of the ecosystem to be managed. Responsiveness to uncertainty takes several forms when it comes to management (Walters 1986, p. 9):

- trial-and-error management, where early management choices are essentially haphazard and later choices are made from a subset that gives better results;
- passive adaptive management, where historical data available are used to construct a single best estimate or model for response, and the decision choice is based on assuming this model is correct;
- active adaptive management, where data available are used to structure a range of alternative response models, and a policy choice is made that reflects some computed balance between short-term performance and the long-term value of knowing which alternative model (if any) is correct.

Active adaptive management has received the strongest endorsement as the way to produce improved management through controlled experiments. Note "experiments" is plural. The chance for improved management is substantially enhanced when the same hypotheses are tested over multiple, but similar units, as "the balance of learning and risks often does not favor experimental disturbances in single, unique, managed systems" (Walters 1986, p. 9). Also note the reluctance to equate active adaptive management to "trial-and-error learning," when the latter little conveys the interactive coupling of learning and management through a proposed process of experimentation, re-experimentation, and continuous hypothesis-generation and testing that guides actual decision making. Lee (1999) also distinguishes adaptive management from trial-and-error learning, though he notes that others continue to equate the two.[11]

For these reasons, active adaptive management is best suited to ecosystems where the human footprint is evident, but not deep; namely, the humanly colonized but not dominated ecosystems. Here is where a series of experiments can be carried out at multiple localities on ways to minimize human disruption and restore or rehabilitate ecological functions and processes, in whole or significant part, to what existed historically or prior to human settlement there. The decision space remains dynamic, but the unintended effects of an experiment on humans are by definition minimized. It is also here where the uncovering of new or unexpected uncertainties in the process of adaptive management experiments can be better tolerated by decision makers, who are not under pressure to come up with the right answer the first time around. As we will see in the next section, passive adaptive management and other forms of "adaptive management," where experiments and restoration may play less of a role than monitoring, assessments, and adjustments, is suitable to more densely utilized systems.

Active adaptive management, nevertheless, has been recommended for urban and agricultural ecosystems, whose defining features are comparatively high population densities and/or extractive uses. Some of the best-known ecosystem management initiatives (see Gunderson et al. 1995), particularly for the Chesapeake Bay, Great Lakes, Rhine River, Louisiana (Mississippi River-Delta), Everglades, Columbia River Basin, and, most recently, the San Francisco Bay–Delta system, incorporate a mix of ecosystems, including agricultural, urban, and less populated ones. While many initiatives explicitly adopt an "adaptive management" approach, much of their management learning does not or will not take place through experimentation, re-experimentation, and hypothesis testing. In fact, the lack of formal experimentation is the chief distinguishing feature of these initiatives (e.g., Johnson et al. 1999). In reality, they take place in zones of conflict between increasing human populations, resource use, and demands for environmental services, where actual resource management and learning have to be tailored to the specific ecosystems and the landscape being managed. Some commentators (e.g., Lee 1999) go so far as to take such conflict as core to the management and conservation of ecosystems.

Case-by-Case Management

Finally, our hypothetical landscape includes ecosystems where human pressures in the form of rapid population growth, increased natural resource utilization for extractive purposes, and rising pressures for a better environment are in conflict. The paradox is strongest here, because important services are derived from these system *and* these services do not override the importance of ecosystem functions. Conflict is always immanent because the ecosystem's inherent unpredictability and the wider demands for high reliability in resources extracted from that ecosystem are often inconsistent and opposed, especially in the absence of mediating mechanisms that reconcile the demands of high reliability management and adaptive management. Chapter 5 details what we believe to be the central mediating mechanism. Below we outline the nature of the conflict and how it gives rise to its own management regime, case-by-case management.

In zones of conflict, ecosystem services and functions have become more decoupled than ever before, giving rise to the need to recouple them. At this point along the gradient, there is widespread understanding that population, resources, and the environment are truly interconnected (tightly coupled). Still, the management of ecosystem services and the management of ecosystem functions have become decoupled from each other (e.g., the Kissimmee River is now channelized), resulting in substantial negative environmental externalities (e.g., Lake Okeechobee is no longer a lake but a managed reservoir, whose overflows have negative effects on the reservoir's estuaries). Because humans are everywhere in these zones, the potentially negative consequences of management interventions loom large in decision makers' decision space. Flips in land-use and ecosystem patterns become more common, with perennials giving way to annuals, grass to shrubs, natives to exotics, creeks to culverts, rivers to channels, and land-use pattern to more fragmented patterns. While all this is going on, enormous disagreement exists in these zones over almost every facet of what it would take to recouple service and function in more positive ways. In a sentence, the paradox confronting ecosystem management has increased dramatically by the time the decision maker arrives in a zone of conflict. Each case of management becomes a battlefield, each consensus or agreement becomes an armistice. In these dynamic circumstances, the decision maker is best advised to manage each case on its own merits.

What is meant by case-by-case management? First, it must be said that a great deal of case-specific analysis and management is already going on. Many ecosystem researchers and decision makers already see themselves treating or trying to treat each case on its own terms. "The best approach is case by case," says ecologist Ted Case, when it comes to analyzing and managing for biological invaders (Enserink 1999a, p. 1836). Similarly, for many of our interviewees, the essence of good management is taking advantage of unexpected, one-off opportunities to promote ecosystem initiatives. A recalcitrant stakeholder is transferred, a new policy window opens with a change in administrations, the right person actually attends the right meeting, a long, haphazard consultation process pays off in unexpected ways—all can be capitalized on, case-by-case. In short, many people already know and accept that any given case differs in important respects from others, where its merits are precisely *its* merits, not those of the others.

Yet, as with the other three management regimes, the argument for case-by-case management is prescriptive, and as such the management regime is more than fine-grained analysis or opportunism. The difference between the latter and case-by-case management is that the case-by-case management is explicitly directed to more effectively recouple services and functions in ways that are tailored to local conditions. Indeed, we define case-by-case management as those occasions when successful recoupling occurs at the field level to meet the twofold management goal. In such instances, the longer term optimization process that drove the initial coupling can be carried on, but frequently in dramatically different (i.e., more case-specific) ways than initially conceived at the policy level. Being opportunistic may thus be necessary for the decision maker, but on its own can lead to taking the ecosystem management down avenues that, while making eminent short-run sense, lose sight of the long-run twofold goal of recoupling improved functions and reliable services more effectively. As seen in chapter 5 and its examples, case-by-case management, as more narrowly defined, is also found in practice.

What, then, are its principal features? First, case-by-case management is evolutionary and necessarily so, in light of the decision space in which decision makers operate and the zones of conflict in which their management is undertaken. Decision makers and managers start with the expectation that the ecosystem is out there waiting to be identified. They realize once in the field that there are problems in delineating salient ecosystem features for the purposes of meeting the twofold management goal. Later they acknowledge that such conflict arises in part because what is out there depends crucially on how the "it" they are looking for is defined (or not defined) in the first place. They then end up understanding better what works best by way of recouplings in any particular dynamically interactive situation depends on customizing the ideal and the practical to meet the specific objectives agreed on in resource management as following in real time from the twofold goal, case-by-case. One lesson is that as ecosystem management projects proceed, objectives change, concluded one of our interviewees. Note that evolution and redefinition often go together. "Like many other government programs, ecosystem-based management is the result of an evolutionary process of experimentation, goal definition and redefinition, and the search for appropriate implementation strategies," in the words of Imperial (1999, p. 460). In such circumstances, it is not surprising that case-by-case management draws from the very different approaches of complex adaptive systems theory, adaptive management, and high reliability theory at different stages of this evolutionary process. Nor is it unexpected that these different approaches are accented and emphasized differently in the evolutionary process, given that all four management regimes are potential alternatives for ecosystem management anywhere in the landscape, as we saw earlier.

Second, case-by-case management means analyzing each case of management on its own terms. At least five different criteria exist to evaluate management: (1) in terms of whether the management achieves its stated objectives; (2) against the ideal recoupling of services and functions, which may or may not match the management objectives; (3) against implementation records of like management; (4) in terms of the counterfactual, that is, what would have happened had not the management been in effect; and (5) according to whether savings could be realized if the management were undertaken more cost effectively. No one criterion is better than another, nor could it be in a decision space where most important decisions are the subject or object of conflict. In this setting, judging each case on its merits is deciding the mix of criteria and weights to be assigned to each criterion for the case in question (for a discussion of multiple evaluative criteria in ecosystem management, see Cortner et al. 1998).

Third, not only do multiple criteria exist to assess management strategies, but the actual implementation of these strategies also results in evaluating what are the appropriate mix and weights for the criteria. Decision makers really do not know how to evaluate effective site-specific operational recouplings of functions and services, until they see just what those recouplings actually are, on the ground and in real time. When this happens, the case is also being analyzed and managed on its own merits. In this way, case-by-case management is locally syncretic rather than globally synoptic. Berry et al. (1998, p. 74) summarize this issue for the Columbia River Basin: "observations call attention to the overarching importance of situation, setting, or context, especially with respect to wholesale changes ecosystem management and the research supporting it demand."

Fourth and importantly, case-by-case management relies on triangulation as the primary method of analysis, confidence-building, and where possible, consensus generation. In a decision space that is dynamically interactive, many, if not most, parties to a conflict, including the experts, are in the grip of many unknowns, frequent surprise, and little agreement. Where few involved know what really is in their best long-run interests, and where most everyone is playing it by ear, including the so-called power brokers, it is impossible to expect one, true, accurate picture of what is going on or should go on to emerge in meeting the twofold management goal. In this decision space, it is best for decision makers to triangulate on what they should be doing. In fact, one does find triangulation recommended for ecosystem management (e.g., see Berry et al. 1998, p. 76, with respect to the Columbia River Basin).

Triangulation is the use of multiple (the "tri" need not refer to just three) methods, databases, theories, disciplines, and/or key informants to converge on what to do about the issue in question. The goal is for decision makers to increase their confidence that no matter from what direction they analyze the issue (i.e., no matter how they describe the problem), they are led to the same initial conditions, specific alternative, or recommendations with respect to the twofold management goal (for more on triangulation, see Denzin 1970; Cook et al. 1985; Brewer and Hunter 1989; Moris and Copestake 1993). To be clear, one possible outcome of a triangulation is that decision makers reach agreement that one of the four management regimes is most appropriate at a given point in time for a specific site within the zone of conflict. At some "point," there simply is no substitute for biodiversity or our need for it. Thus, it is possible to find area-specific adaptive management or self-sustaining management being undertaken within a zone of conflict, but always on a case-by-case basis. However, for the whole zone of conflict the challenge is to meld together a variety of management strategies.

Ideally then, the instruments used in triangulation should be as diametrically different (indeed, orthogonal) as possible. "Convergent findings are compelling only if it can be demonstrated empirically that when the methods err, they typically err in the opposite ways" (Brewer and Hunter 1989, p. 18). From this standpoint, the conventional debates between "quantitative versus qualitative," "reductionistic versus holistic," and "positivist versus postpositivist" approaches to analysis miss the point altogether. In a tightly coupled, dynamically interactive decision space, the decision maker needs all of the methods—and the more different they are, the better to triangulate on and manage issues more confidently. By the same token, the professional polarizations in ecosystem management—ecologists versus engineers, fisheries biologists versus line operators, agriculture versus the environment, local communities versus national environmental organizations, insiders versus outsiders, adaptive management versus high reliability management, ESA versus flexible management—miss the point that when the principals in this polarization actually do converge on a joint position, confidence increases dramatically that the convergence is a good one to pursue. In case it needs saying, there is never any guarantee that triangulation will occur or, if it does, that the convergence will have management relevance. The latter, again, depends on the political, social, and cultural values that serve as the initial conditions for triangulation.

Case-by-case management's focus on triangulation and confidence contrasts with its analogues in the three other management regimes. At the risk of simplification, high reliability management is centered around clear causal understanding and certainty over key management technologies. Active adaptive management is just as clearly preoccupied with falsification of hypotheses, when not focused on uncertainty identification and reduction (in fact, adaptive management may identify new uncertainties relevant to ecosystem management).[12] Self-sustaining management, in contrast, is much more "accepting" of causal uncertainty, since out of the complex interactions of the ecosystem comes self-organization, which cannot be created through the other forms of management.[13]

These different foci have profound consequences for the role of stakeholder involvement in each of the management regimes. All regimes share the same first-order stakeholder involvement issues in that culture, society, and politics enable or restrict broad-based participation in setting the terms of reference for all aspects of ecosystem management. Second-order issues are, however, different across management regimes. Principally, the focus of case-by-case management on triangulation and confidence building in the face of a conflicted, dynamic decision space calls for a much wider definition and continuous involvement of stakeholders than the other three regimes. Indeed, the call for a more open stakeholder process in adaptive management and ecosystem management (e.g., Schlinder and Cheek 1999) is perhaps the best indirect measure we have that the management is taking place in a zone of conflict requiring case-by-case management. For adaptive management and certainly for self-sustaining ecosystems, the second-order stakeholder issues are driven more by ensuring involvement of experts, be they engineers, ecologists, or decision makers themselves (e.g., Daily et al. 1996). With a wider array of stakeholders, case-by-case management also takes on a wider set of defined and redefined services and functions important in ecosystem management. We return to these stakeholder differences below.

The features of case-by-case management—its evolutionary nature, the importance both of multiple evaluative criteria and implementation, and the key role of triangulation and diverse stakeholder involvement—mean that there are many "models" at work in case-by-case management compared to the fewer individual-based models of complex adaptive systems theory and the more formal ecological models being developed and tested through adaptive management. Case-by-case management draws from these approaches as well as from the informal, tacit knowledge/bounded rationality models of decision makers themselves. (Indeed, the five evaluative criteria constitute their own "models.") This is why "passive adaptive management" is so often case-by-case management, as it involves a broader range of modeling than does active (i.e., experiment-based) adaptive management.

These features also mean that case-by-case management in zones of conflict is rarely a total failure or a total success when it comes to the operational recouplings needed to meet the twofold management goal. The more evaluative criteria for site-specific management, the longer the evolution period of the management, the more different the information to be assimilated and the more stakeholders involved, the greater the chance that management will be analyzed and assessed in favorable terms on some, but not all, the criteria. This way, the performance record of case-by-case management will almost always be mixed. Neither will it ever be totally negative nor entirely positive, which is why it is always difficult to generalize or replicate from case-by-case management.

Ecosystem Management Over the Four Regimes

Summary and Discussion

Table 4.3 summarizes the chief features of self-sustaining, (active) adaptive, case-by-case, and high reliability management regimes. Overall, table 4.3

Table 4.3 Distinguishing Characteristics of Management Regimes

	Self-sustaining management	(Active) adaptive management	Case-by-case management	High reliability management
Ecosystem properties	Relatively untouched by human hands	Limited human disruption of essentially natural ecosystem	Zones of conflict with natural systems overlaid by human artifacts	High human populpation densities and extraction define the ecosystem
Beliefs about ecosystem being managed	Sustainability is assured through natural, complex processes, such that the ecosystem manages itself optimally	Ecosystem can be incrementally managed over time to mimic its presettlement (historic) functions and processes	Each ecosystem is a case unto itself, where managers improve ecosystem performance, but without full success	Resource system is predictable within known tolerances and can be managed safely for peakload performance
Types and foci of models employed	Ecosystem builds and uses its own internal models and exhibits behavior that defies extensive formal modeling. Focus is on self-organization arising out of dynamic processes	Formal theoretical models and hypotheses are tested as a learning process. The focus is on falsifying hypotheses, reducing uncertainty, and better identifying core uncertainties	Variety of formal and informal models are used and weighted differently reflecting merits of each case. Focus is on increasing confidence through triangulation	Extensive thought experiments or computer modeling and simulations are used before scaling up to real-time management. Focus is on clear causal certainty over key management technologies
Modes of learning	Learning is in the natural systems as part of its evolution, not in human beings	Learning-by-doing, that is, through management experiments, preferably large-scale and multi-site	Learning through real-time management and evaluation of treating each case in its own right	Simulation, experimentation, and search for improvement while unable and unwilling to test the boundaries of reliability
Stakeholder involvement	It is key in selecting and maintaining ecosystems for preservation	It is key in generating hypotheses and strategies and in implementing and evaluating experiments	It is key to providing input to the triangulation for obtaining a large variety of perspectives	It is key to defining services and providing oversight on service reliability
Measure of success	System is successful as long as it is self-sustaining	Science-based learning leads to improved understanding, and this leads to better management	Success measured against multiple criteria, such that it is always mixed	Success is meeting peakload demands safely, all the time
Measure of failure	Failure is human intervention that disrupts or otherwise alters an ecosystem that is already self-sustaining	In the long run, there is no "failure," as long as learning for better management takes place in light of experimental results	Failure is inevitable to some degree because of multiple criteria, that is, a success cannot be generalized beyond the case at hand	Failure is when a HRO does not meet peakload requirement safely on its own terms, that is,when no "act of God" or exogenous factor caused the failure

Source: adapted from Gratzinger in Roe et al. (1999).

clarifies an earlier point: the unrestricted and undifferentiated preference of one management regime over another—be it self-sustaining, adaptive, case-by-case, or high reliability management—across all ecosystems within a highly variegated, variously populated, and diversely utilized landscape is not only inappropriate, it is lethal to the goals of effective ecosystem management. Meeting the twofold management goal requires tailoring management to the ecosystem. Such customization must be done in ways that take full account of the differences between ecosystems arising not only from their internal dynamics but also from humans' changing demands along the gradient placed on ecosystem functions and services across time and space. This conclusion has profound implications for the tension between adaptive and high reliability management approaches.

The common view, as already discussed, is that adaptive management has the unenviable task of trying to correct for the high reliability management of natural resources in the past. From the framework's perspective, it would be invidious to the goals of effective management to rely primarily on active adaptive management in zones of conflict, where high reliability management is a priority as well:

- While safety (e.g., in fire management, levee protection, entrained water supply) may well be of concern to ecologists and other scientists as adaptive managers, in no way have these experts been recruited, trained, indoctrinated, and continually monitored to ensure that safety and reliability are their number-one priority.
- While ecologists as adaptive managers may have some management skills, in no way have they been professionally trained to be technically competent, high performing managers committed to producing, at the same time, ecosystem goods and services and their safe provision.
- While ecologists in adaptive management may at times see their management as intrinsically tied to politics and a turbulent task environment, they have almost all been trained to separate science from politics, the technical from the social, and the organizational from the technological—a set of distinctions incompatible to high service reliability.
- While ecologists may well be professionally committed to adaptive management, they may not live and work in the areas where it is to be undertaken and where impacts of possible error are felt most acutely.
- While ecologists as adaptive managers understand the risks associated with trial-and-error learning, they all have been trained to see the primary virtue of adaptive management as the ability to reject hypotheses, that is, to risk and actually accept failure as a way to learn about the system of interest. Where high service reliability seeks to avoid failure at all costs, adaptive management welcomes it because of the learning that it entails.

Case-by-Case Management: Threshold or Regime?

The factors listed above do not automatically mean that adaptive and high reliability management regimes are incompatible everywhere. "Adaptive organizations," according to Grumbine (1997, p. 46), "construct networks for information sharing, train and encourage messengers, reward bridge builders, and welcome new learning"—a description that could also be applied to HROs. In fact, this book conceives case-by-case management as a transition between adaptive and high reliability management regimes, that is, an adaptive management directed more to deriving reliable services from improved ecosystem functions, and a high reliability management directed more to improving the functions from which its services are derived. In zones of conflict, case-by-case management may not be a full-blown regime, but in fact the threshold between adaptive management and high reliability management. Or it may only be a transitional phase while moving to the left or the right of the gradient. We argue, however, that there are many ecosystems in zones of conflict for the foreseeable future, where moves to the left or the right of the gradient are for all intents and purposes impossible. This merits developing case-by-case management as a full-blown regime with its own theory and practice. No doubt the ecosystems in zones of conflict are in need of better recoupling of improved functions and reliable services. We undertake one such crosswalk between adaptive and high reliability management regimes in chapters 5 and 6.

Shifts Between Management Regimes: Continuous and Discontinuous

Our key policy question of how to manage is best answered from the perspective of table 4.3 by requiring decision makers to recognize the learning-based thresholds across the gradient. These threshold were discussed earlier, but only in general terms. What are they specifically? Assuredly, empirical "triggers" are involved, observed and learned about (e.g., we learn when different kinds of population densities lead to different kinds of land use changes). The triggers of change in ecosystem management, according to our framework, are what decision makers have learned to be the limits of what they can optimally manage under the model of learning that governs the management they are then undertaking. Some of the limits of this learning, such as the adherence to the precautionary principle, have been touched upon. More generally, work by Louise Comfort (1999) suggests some characteristics of these learning-based thresholds: they are highly sensitive to the initial conditions of the landscape and ecosystem. They are defined by their vulnerability to outside forces (such as fire, earthquake, or other emergency). They are irreversible in important respects for management (i.e., once decision makers continue down the road of one kind of management, it is hard to turn back). They drive decision makers' mutual adjustments, which become self-reinforcing over time and eventually become rigidities that force decision makers to self-organize around different ways of learning and management regimes. Finally, they are the source of unpredictable impacts and of predictable patterns of behavior that seek to reproduce themselves within the ecosystem and extend themselves, inappropriately, to other ecosystems. In brief, thresholds are those transitional phases when people in the landscape or ecosystem "discover" a common threat to their current ways of learning-based management and, in response, organize alternative ways or "states" of learning and management along the approaches just outlined. (For a range of ecology perspectives on state-and-transition models, see Westoby et al. 1989).

How does this actually happen? Consider the three internal thresholds in figure 4.1 and what they imply. Starting on the left and moving right, the shiftpoints are when people in the landscape/ecosystem have learned (1) they can no longer rely on biological evolution (as in complex adaptive systems theory) to make the required ecosystem changes; (2) they can no longer rely on experimental learning (as in adaptive management) to make the required ecosystem changes; and (3) they can no longer rely on case-specific triangulation (as in case-by-case management) in making the required ecosystem changes, at least in the continuously highly reliable fashion now demanded.

For example, over time, as people in a region watched elephant numbers decline from hundreds to fewer than ten or so, they went from letting the elephant habitat manage itself, to trying to manage the agro-ecosystem adaptively through innovative interventions and management improvements, to undertaking more urgent interventions that their increasingly specific and unique conditions called for (in some cases, anti-poaching measures, in other cases, restricting the incursion of agriculturists or stopping deforestation), and finally to stationing a guard with each of the remaining elephants so as to reliably ensure its ongoing safety and survival. None of these changes need have been marked by milestones where specific numbers of elephants triggered one mode of management over another. What triggers change is what resource managers have learned to be the limits of what they can satisfactorily manage while operating under the model of learning associated with the management regime under which they are operating.

As noted earlier, these thresholds have both discontinuous and continuous features, as the gradient's dimensions themselves are characterized by continuities and discontinuities. Population growth, increasing extraction, the demand for discrete ecosystem resources and services, and the shift to increasingly organizational interventions to manage ecosystems all have the potential to transform ecosystems gradually or by flips. Furthermore, the transitions between the management regimes are also marked by continuities and discontinuities as also illustrated in figure 4.3. There are flips from one management to the next, but for each ecosystem category only one of the four management regimes is optimal. Because the four regimes are present at any given time, shifts from one dominant regime to the next might actually be more gradual. It is accumulated learning about ecosystems, the limits to any one way of learning, and the push and pull of new ways of learning, that forestall thinking of each ecosystem as so dynamically unique that it cannot really be compared to other ecosystems, save for some basic set of functions (on the uniqueness of ecosystems, see Slocombe, 1998, p. 488). Ecosystems are unique, but not incommensurable as long as learning takes place across the gradient.

Where Similarities Matter More than Differences

In the same fashion, for all the differences between management regimes, the similarities and complementarities must also be stressed. At first glance, the "light impact" management recommended by complex adaptive systems theory for relatively unmolested ecosystems stands in sharp contrast to the decidedly "heavy impact" management recommended by high reliability theory for highly utilized ecosystems. The latter asserts that human reason can fully design, manage, and control a complex technology, except for forces of nature, such as a natural disaster. The former asserts that natural forces can design, manage, and control a complex ecosystem, excepting for human interventions, which cause human-made disasters. Indeed, nothing could seem farther apart than high reliability management that has systematically devastated self-sustaining ecosystems and self-sustaining management that could have preserved these ecosystems, if left on its own.

Yet on closer inspection, self-sustaining and high reliability management regimes have several important features in common. High reliability organizations build redundancies into their technologies and operations; ecosystems have suites of organisms with equivalent functions, where if one type is removed it is replaced by another in the same functional group (Stone 1995). This engineering–ecology connection is made explicit by Shahid Naeem, an ecologist, in his "Species Redundancy and Ecosystem Reliability" (1998, p. 39):

> a central tenet of reliability engineering is that reliability always increases as redundant components are added to a system, a principle that directly supports redundant species as guarantors of reliable ecosystem functioning. I argue that we should embrace species redundancy and perceive redundancy as a critical feature of ecosystems which must be preserved if ecosystems are to function reliably and provide us with goods and services.

Other ecologists have made the same or a similar point. "Lack of apparent species 'redundancy' within functional groups is a dangerous condition," according to ecologist Brian Walker.[14] Others see protecting major ecosystems explicitly in terms of redundancy, as species with overlapping functions (Peterson et al. 1998, p. 9) and as buffers against species and ecosystem losses elsewhere (Clark 1996, p. 112; Folke et al. 1996, p. 1020). Ecosystem resilience is seen in just such terms as well. That is, as the ecosystem's capacity to buffer disturbances (Folke et al. 1996, p. 1020). In fact, organization theorist Paul Schulman has drawn on the work of ecologists and others in equating high reliability in organizations to a kind of resilience (e.g., Schulman 1993b). For Grumbine (1994, p. 35), ecosystem management itself "means comprehending the balance between core reserves, buffers, and the matrix of lands used more intensively by humans." The language of ecologists sometimes even takes on a high reliability tint, not just in Naeem's work but also in that of others who, for example, refer to keystone species as "ecosystem engineers" (Folke et al. 1996, p. 1019; Callicott et al. 1999, p. 31).[15] Sometimes redundancy is deliberately built into a project for recovery or restoration reasons. In one CALFED gaming exercise, the "biological bar" was raised so that the baseline water requirements that had to be met first through allocations from the Environmental Water Account included those that assured fisheries agencies that every possible contingency could be met to ensure safety of a particular species.

The Wraparound of High Reliability and Self-sustaining Management

Another similarity between management regimes stems from HROs' insistence that reliability is, after a point, nonfungible. In high reliability management, safety is such a priority that it cannot be traded for money or other services; at some point, there is no substitute for reliability and the safety that comes with it. So too is the desire to protect and preserve ecosystems motivated by the need to secure their safety reliably over time. Thus, the stewardship infrastructure that has to be in place to preserve existing (nearly) self-sustaining ecosystems is often daunting in its own high reliability requirements. Whole bureaucracies are dedicated to maintaining such infrastructure. In so doing, they often rely on the very same sophisticated management technologies already described as high reliability in the literature, for example, emergency control centers, fire-fighting units, and air traffic control (e.g., see Harte 1996, p. 28).

In this way, the protection and preservation of self-sustaining ecosystems through high reliability management are very expensive and demanding enterprises in terms of infrastructural, technological, and financial resources. Obviously, adaptive management and case-by-case management also require such resources, but these management regimes differ from the other two in that both adaptive management and case-by-case management explicitly *accept failure* and mixed performance records. Contrast that to the high reliability management of the last few remaining, relatively unmolested ecosystems, where failure is not an option for the decision maker and where a mixed performance record cannot be risked. In the latter two management regimes, any casualty due to human error must be avoided. Protecting the system reliably and safely is the first priority, all the time, because there are too few of these ecosystems left. There is only one Everglades, one Columbia River Basin, one San Francisco Bay–Delta. Lose one and it is irreplaceable. The stakes are most clear and highest in self-sustaining and high reliability management regimes, and accordingly their requirements for public support and resources are quite large.

This wraparound from high reliability management back to self-sustaining management (see also the discussion on figure 4.3) is, however, not without profound effects, for it continues the process of redefining ecosystem services and functions away from the presettlement template, this time irrevocably. At the point where it becomes necessary to put stewardship infrastructure in place to protect an ecosystem, it is no longer the original ecosystem that is being preserved. As we noted in chapter 3, the Everglades is on nothing less than a life support system of control room, pumps, canals, and facilities around "its" perimeter. Without the sophisticated technology that now protects and manages the Everglades, and it would revert to a more "natural state," but it would not be the Everglades as we know it.

The high reliability management of internally self-sustaining ecosystems means that the ecosystem functions associated with the latter are now being recoupled with those ecosystem services (re)defined by the former. Thus, what once was considered an offensive question impossible to answer— "How do you price a unique ecosystem?"—becomes in the wraparound a question with an all-too-often obvious answer: it costs a lot. It is next to impossible to monetize self-sustaining functions on their own, but monetization is inevitable and taking place all the time, if only in the form of opportunity costs faced by the HROs responsible for ecosystem management interventions (e.g., BPA, SFWMD, and the DWR).

For many ecologists and others, some variant of the wraparound is inevitable. However, as one senior scientist at the Nature Conservancy said, protection and preservation are simply not enough. Saving what is left leaves only nature's cemetery (Luke 1997). Much more needs to be done to restore and rehabilitate ecosystems, albeit there is great conflict over just how this should be done. The conflict puts us back into the case-by-case management regime, if we are to come up with new ways to promote restoration. Laying out one such alternative is the task of next two chapters.

5 Ecosystems in Zones of Conflict

Partial Responses as an Emerging Management Regime

Many of the most expensive and important ecosystem management initiatives under way today are in "zones of conflict" between increasing human populations, resource utilization, and demands for environmental services. The four cases in this book—the San Francisco Bay–Delta, the Everglades, the Columbia River Basin, and the Green Heart of the western Netherlands—are no exception. Each combines the need for large-scale ecosystem restoration with the widespread provision of reliable ecosystem services. As seen in chapter 4, case-by-case management is the regime most suited for such contentious issues in zones of conflict.

It is no small irony, therefore, that these ecosystem management initiatives are often presented as showcases for adaptive management (e.g., Gunderson et al. 1995; Johnson et al. 1999). This showcasing is understandable when we realize that here the paradox is at its sharpest. Consequently, the initiatives are unique in the considerable amount of resources made available to adaptive management or ecosystem management, precisely because the ecosystems are in zones on conflict. Much of the funds come not from natural resource or regulatory agencies, but from the organizations that produce and deliver services from these ecosystems, such as water-supply or power-generation companies. In southern Florida, the Army Corps of Engineers (ACE) and the South Florida Water Management District (SFMWD) estimate the costs of their proposed ecosystem restoration plan to be $7.8 billion; in the Bay–Delta, the CALFED Program expects to spend about $10 billion during this implementation having already spent more than $300 million on ecosystem restoration in recent years; and in the Columbia River Basin, the Bonneville Power Administration (BPA) alone provides some $427 million per year for fish and wildlife measures. As a senior BPA planner remarked, "We are the largest fish and wildlife agency in the world." Contrast these millions and billions to the funding problems often reported by "purer" forms of adaptive management for ecosystems towards the left of the gradient in figure 4.1.

In short, although important services are derived from these ecosystems, the services do not override ecosystem functions, thus raising the resource demands of ecosystem management. When we move to the right of the gradient toward high reliability management, services dominate functions, thereby putting less pressure on service-oriented and other agencies to provide resources for ecosystem management. The same happens when we move to the left toward adaptive management: few services are provided and thus the overall need for reliability decreases.

Seen in this way, there are several reasons why ecosystems in zones of conflict are important and why there is a high premium on proposals to help deal with them. First, zones of conflict are the least understood category in terms of management implications, while many of the important ecosystems decision makers are trying to manage fall into this category—such as the cases in this book. Second, apart from our framework, there are no proposals or conceptual frameworks available today that guide decision makers in dealing with the tensions and synergies between high reliability management and adaptive management. Third, in the absence of such proposals, the pressure is to pull the ecosystem to the left, giving priority to functions (i.e., "reduce extraction and population densities") or to the right of the gradient, giving priority to services (i.e., "manage the ecosystem resources needed for reliable provision of services, such as quality water"). Fourth, the demand for ecosystem services in these areas not only threatens successful ecosystem management, it also offers opportunities, such as the availability of financial resources, as long as decision makers can successfully recouple more improved functions and reliable services.

The Paradox and the Coupling–Decoupling–Recoupling Dynamic

Just what does case-by-case management actually look like for zones of conflict? There are many ways to define it. Whatever form it takes, it has to address the core impasse between adaptive management and high reliability management. The next chapter presents a proposal for bandwidth management that aims to address the paradox and provide a management regime for zones of conflict.

The challenge is to develop a crosswalk between adaptive management and high reliability management that enables meeting the twofold management goal of recoupling ecosystems functions and services. To do so means connecting the regimes so as to make them mutually reinforcing, instead of being barriers to each other's success. The partial responses described in chapter 3 show that this is what agencies in the case studies are already trying to do, notwithstanding perceptions to the contrary. The fact that a program like CALFED seeks to prioritize and underwrite adaptive management across the board—even for its high reliability components—obscures the reality of their attempts to reconcile the need for adaptive management with their equally important, if not more important, high reliability mandates. The agencies in the initiatives are, in fact, inventing a new management regime: case-by-case management. Thus, we must reinterpret our interviewees' much-repeated complaint and apology that the practice of adaptive management lags behind the theory. Instead, what is happening instead is that *the managers in the case studies are ahead of the literature in inventing case-by-case management.* Not only have their innovations gone unrecognized in the ecosystem management literature, the literature and even the decision makers themselves continue to judge these initiatives by the inappropriate standards of adaptive management.

Before we can turn to an in-depth discussion of the partial responses, we must introduce the coupling–decoupling–recoupling dynamic (CDR dynamic), a major organizational mechanism that we found operating in the case studies and from which a crosswalk can be produced. The CDR dynamic helps us make sense in locating and maintaining the effectiveness of the more important innovations and proposals found in the case studies (chapter 3).

Coupling

At the policy level, there is broad recognition of the need to recouple functions and services in each of the cases. We saw the distinctly interconnected nature of the subsystems of interest. Decision makers confront many physical, biological, chemical, social, and political links between issues of water quality, water supply, protection of endangered species, ecosystem restoration, hydropower generation, and flood control. In fact, what drives major ecosystem management initiatives is this necessity of addressing these and related matters simultaneously. It follows that policies and programs dealing with interlocked issues jeopardize their effectiveness and further threaten the ecosystem if they are themselves not coupled in important respects.

Decoupling

Yet attempts at policy coupling generate their own counterforces. Interlocked issues, as we will see, become unmanageable when different and already complex policies are treated as being just as tightly coupled as the world they seek to change. The initially valid recognition that issues are so interrelated that they have to be optimized together ends up rendering policy difficult if not impossible to achieve in these terms alone. When faced with such a turbulent task environment, the pressure is, as any number of organization theorists continue to underscore, to decouple the issues of specific interest from that environment and buffer them in the form of their own programs, agencies, or distinct professions (e.g., Chambers 1988;

Hukkinen 1999). The preceding chapters have reiterated how this organizational fragmentation characterizes ecosystem management as well.

Decoupling at the program, agency or professional level is most apparent when tradeoffs are impossible or extremely difficult to make. A Bonneville Power Administration biologist told us that "flood control always wins," because the Army Corps of Engineers sets the water schedules, and its primary mandate is flood control. The power administration then optimizes for power and fish within these schedules. In doing so, the Corps has in essence decoupled flood control from power and fish by posing the former a fixed constraint for real-time operations, instead of as one of the objectives against which the operators can trade off. Another oft-repeated example is the Endangered Species Act (ESA) standards that have to be met, where the costs of their enforcement are not an issue in the view of the regulatory agencies. Endangered species have to be saved, whatever the expense; emergencies are never cost-effective. In this way, the standards decouple species protection from reliable water supply and quality, where cost indeed is always an issue. The 1999 delta smelt crisis in the Bay–Delta (chapter 3) shows just how far-reaching can be the consequences and cascading effects of the inability to strike tradeoffs.

Thomas (1997, p. 221) observes similar decoupling mechanisms in interagency cooperation on ecological problems in California: "Ecologists accept interdependence among public agencies, and even welcome it, while agency executives generally seek autonomy from one another in order to provide stability and certainty for their organizational units." What is coupled at the policy level, becomes decoupled at the program, agency, or professional level for implementation or management. Thus, the decision maker sees all manner of population, resource, and environmental programs professing their connectedness, but in the real world these programs operate on their own, with professionals often trained in separate disciplines.

The decoupling, while achieving short-term reductions in turbulence and increases in program stability and effectiveness, ends up undermining the very optimization process that drove the initial systemwide coupling. Decoupling ultimately serves to emphasize the interdependencies of the agency's goals and effectiveness with those of other actors in its environment. The interdependencies can take many forms, ranging from experiencing the effects of related programs in other agencies to direct interventions in the internal operations under the Endangered Species Act, as described by Thomas (1997, pp. 242–243):

> Faced with the very real possibility of losing broad decisionmaking discretion and management autonomy to the narrow cause of species protection, the directors [of resource management agencies] turned to staff ecologists to develop plants to manage the habitat of listed (and potentially listed) species to maintain viable populations of these species before the agencies could be sued under the Endangered Species Act.

The result was increased interagency cooperation. In brief, decoupling serves to highlight how interlinked the issues really are and how important it is to deal with them in a direct, coupled way.

Recoupling

Where the initial coupling generated pressure to decouple, the decoupling reinforces the pressure to recouple—the third element of the CDR dynamic. The interconnectedness of population, resources, and the environment has to be reflected in policy. The twofold management goal can only be met if the couplings are not only at the policy level but actually operationalized "corner to corner." In particular, we argue, it must incorporate line operators—the people who actually manage the various activities in the field in real time—at every stage of policy development, implementation, redesign, and ongoing management. The lesson of the implementation literature is that operations are never "just operations" but are de facto policymaking, at times even more important than formal policymaking. A crucial characteristic of this operationalization is that it enables a dynamic optimization process among different (often interagency) goals and issues, thereby making program and agency boundaries permeable. "Dynamic" means that the "variables" of the optimization problem can be manipulated at the same time to explore tradeoffs and priorities as well as capitalize on opportunities for flexibility. We will see examples of recoupling later this chapter. The decoupled situation, in contrast, supports a static optimization process at best, where a program or goals set fixed boundary conditions within which line operators optimize the effectiveness of the program for which they are responsible, for example, water supply operators working within Endangered Species Act mandates.

How this operationalization actually occurs is necessarily case-by-case, since the dynamics of coupling, decoupling, and recoupling are inevitably site-specific and contingent on factors specific to the situation at hand over an already differentiated landscape. The longer term optimization process will be carried on, but frequently in dramatically different (i.e., more case-specific) ways than initially conceived at the policy level. Should it need saying, a latent function of successful recoupling (successful in terms of the twofold management goal) can be to leave the preexisting programmatic, agency-wide, and professional decoupling in place, albeit more permeable in places than before.

A Model of Policy Formulation and Implementation

The CDR dynamic can be taken as different phases of policy formulation and implementation. The case material show instances that conform to an image of three sequential phases: first there is policy coupling, then programmatic decoupling, and then operational recoupling. Further, our proposal for ecosystem management in zones of conflict will be directed toward recoupling, which uses the dynamic as a model of policy formulation. Nothing, however, is automatic or guaranteed about progress toward recoupling as the endpoint.

Empirical evidence shows a great deal of shuttling back and forth between the initial policy coupling and the programmatic decoupling, where the latter is seen as a failure that is then countered at the policy level. The alternative is its acceptance as a necessary and productive step to build on in establishing operational recoupling. Also, some recouplings require intensive effort to sustain them and therefore might ultimately break up, reverting the system back to a decoupled state. In another scenario, the recoupling produces learning feedback to the policy level and revised policies—much as implementation can inform policy priorities.

In using the dynamic as a model of policy formulation and implementation, we must avoid misinterpreting decoupling. When decision makers work to recouple operations in ways that meet the twofold management goal, it is tempting to read decoupling mainly as a barrier to achieve the desired goal. This misses the positive features of decoupling, to which we now turn.

A Positive Theory of Decoupling

The endpoint of ecosystem management initiatives is certainly not to achieve decoupling. Quite the reverse is true in light of the twofold management goal. On the other hand, decoupling is Janus-faced in that its demerits are matched by merits. Not recognizing these merits—in other words, not understanding the "good reasons" why decoupling persists—leads to recoupling attempts that are at best only moderately effective and at worst harmful.

One way to identify the positive features of decoupling is to look at the assumptions behind its negative effects. When Berry et al. (1998, pp. 61–70) write that the Interior Columbia Basin Ecosystem Management Project (ICBEMP) initiatives are plagued by "polarization and fragmentation," "confusion," "coordination problems," and that "no one is either in charge or accountable," that it is "a system out of control," it is easy to see the assumptions behind their assessment. Just as any question assumes knowledge of what would constitute an answer to that question, so too the authors' long list of problems and complaints implies their ideal solution: a well-coordinated arrangement with clear lines of authority and responsibility, in short, an effective hierarchy. Predictably, then, they recommend a "single organization to manage and coordinate ecosystem management research in the Pacific Northwest" (Berry et al. 1998, p. 75).

Without downplaying the negative role of decoupling, many of the "problems" Berry et al. and others describe have a positive role as well. Imperial (1999, p. 458) argues that "the research does not support the proposition that centralized, hierarchical arrangements are superior to ... polycentric arrangement," where polycentric arrangements are the networks of multiple, relatively autonomous agencies currently involved in ecosystem management. Criticism like that expressed above "fails to consider the full range of transaction costs [associated with] centralized arrangements." Building on Imperial's analysis of institutions for ecosystem-based management, the positive features of decoupling can be identified and extended.

The left-hand column of table 5.1 lists the recurrent problems of decoupling. Many agencies, including those mandated to ensure service reliability, have ecosystem management elements in their mandates. This duplication, of course, creates and supports overlap in staff, projects, research, and expertise; all of which indeed utilize more resources. Nevertheless, given the complexity of the system being managed or intervened in, the duplication is also much-needed redundancy for the reliability of large-scale organizations (see Landau 1969; Lerner 1986).[1] This and other positive features are listed in the table's right-hand column.

Table 5.1 Theories of Decoupling

Feature	Negative	Positive
Duplication	Wasteful overlap	Redundancy
Fragmentation	Segmentation of authority	Functional specialization, economies of scale
Conflict	Turf battles	Guaranteed consideration of different interests, constructive debate over competing proposals
Polarization	Inability to speak with one voice to stakeholders	Transparency of issues, keeping tradeoffs in the public arena
Unintegrated priorities and goals	No systemwide priorities and tradeoffs	Institutional protection of vulnerable goals and interests
Accountability	No accountability for overall performance	Accountability directly related to specific tasks
Disjointed information	Lack of coordination in information gathering and assessment	More error correction, less distortion in aggregating or assessing information
No comprehensive approach	Loss of problem-solving focus and capacity	Decreased turbulence in immediate task environment
Policy change	No one in charge, system out of control, unable to achieve necessary fundamental change	Incrementalist, goal-seeking change
Complexity	Disabling complexity (scatterplot of observations)	Enabling complexity (meta-analysis of results)

Fragmented mandates increase the dependency on interagency cooperation and ensure that many officials review the same projects or policies—a practice that can be cumbersome but which allows for functional specialization and economies of scale.[2] Should a single agency have the unencumbered mandate to review an entire ecosystem management project, it would have to develop all areas of expertise now provided by the set of specialized programs and staff. Conflict has long been recognized for its functional value in socio-economic systems. While turf battles are sure to make all major ecosystem management initiatives difficult undertakings, they also provide the best possible guarantee that different views and interests—as long as they are institutionalized—are taken into account in the design of solutions. Decoupling is also associated with polarization, which goes beyond conflict in the sense that the divisions run so deep that they are apparent to outsiders. Government is itself divided, that is, it does not speak with one voice, as in the case described by a HRO executive officer we interviewed: "All the agencies are at loggerheads, instead of cooperating. When the Clinton Administration saw the chaos, the fighting between the agencies, they concluded it was not good for public policy and also that all the agencies should speak with one voice. The effect has been to muzzle any one individual speaking out in a leadership role. Our [head] has been under a gag order since then." Put in these terms, the positive role of decoupling clearly is to keep polarized issues in the public arena. Decoupling increases the transparency of those tradeoffs not made inside programs or agencies that may seek to fold or otherwise obfuscate conflicting issues under one mandate.

Having competing goals embedded in different programs obstructs identifying tradeoffs and setting priorities among them during implementation, especially those that translate directly into issues of authority and funds of agencies. Coupling the goals through one program, however, may not bring striking the tradeoffs any closer, particularly if some goals are more firmly institutionalized than others. A prime example of the latter in ecosystem management is the goal of long-term ecosystem sustainability. As Hukkinen (1999) documented, many agencies have incorporated sustainability in their programs, but in the process of day-to-day tradeoffs among competing priorities, that is, between short-term environmental management and long-term sustainability, the latter routinely fails to leave any impression beyond formal paperwork. The goal is vulnerable because of the absence of organizational and institutional units to develop, promote, and secure it. Such units would then have to be independent, because sustainability must be its own mandate independent of the legitimate, inevitable tradeoffs between this goal and others, including short-term objectives. In other words, for recoupling to work, vulnerable goals, such as sustainability, must be decoupled organizationally. The issue of accountability can be argued along the same lines. True enough, in a decoupled state, no one is directly accountable for the whole. That is, it is beyond any one agency's mandate to make ecosystem management work. Establishing an accountability structure for overall performance may address some issues, but will undoubtedly increase the distance—loosen the ties—between accountability and specific tasks and programs. The latter have clearer, more tangible performance measures from which an organization derives more direct feedback and learning, different from the aggregate measures of success for overall accountability.

The decoupled nature of information gathering often returns as a topic of concern in ecosystem management initiatives. As of March 1999, there were nearly 625 monitoring programs related to the CALFED Program components, with more than 325 devoted to environmental restoration efforts alone. Unsurprisingly and for good reason, CALFED is setting up a more coordinated effort, the Comprehensive Monitoring, Assessment, and Research Program (CMARP). Notwithstanding the obvious advantages of coordinating information gathering, such efforts can be prone to error and systematic distortions as well as making it more difficult to act upon the information (Lindblom 1990; Imperial 1999). Decoupled programs provide more checks and balances against distortion and bring the information user and the information gatherer closer together. The real benefits of any monitoring program, coordinated or decoupled, is always measured by how effective the connection is between information gatherers and users: ideally both should be the same persons (Feldman and March 1981).

The lack of a comprehensive approach to linked problems was cast earlier as the lack of a single, coherent super-agency to undertake an ecosystem management initiative. Yet the loss of problem-solving focus and capacity caused by programmatic or multi-agency decoupling also means decreased turbulence in each program's or agency's task environment. Integrated policy is a disintegrating task environment, unless programs and agencies have mandates that they can address on their own (Leonard 1984). In related fashion, policy change in a decoupled state may not be synoptic or the systemwide shift in policies advocated by ecologists and others (chapter 2). It does, however, better enable incremental change, which is not only a more realistic model of change, but, according to well-known policy experts, also the more rational one (Wildavsky 1979; Lindblom 1990).

In sum, the negative view of decoupling sees it as unnecessary complexity which disables attempts to treat ecological problems seriously; particularly because political, social, and cultural factors limit decision makers' control of the process of learning about the ecosystems to be managed. This is the complexity that turns the experiment into a scatter plot of observations without clear trends. The positive role of decoupling is to focus on the hundreds or even thousands of experiments out there called interventions. Many policy fields, such as unemployment or poverty, have adapted to complexity not by better controlling formal experiments, but by learning from meta-analyses of multiple cases that constitute actual interventions in the real world.

The positive theory of decoupling helps us use the CDR dynamic to answer why current attempts to build a crosswalk between adaptive management and high reliability management are so difficult and fall far short of what is needed to realize the twofold management goal. The answer in short is because, in different ways, they reinforce decoupling, notwithstanding their goal of recoupling services and functions. We explain how interventions might be designed more effectively, after returning to the partial responses found in the case studies.

Partial Responses Revisited

Forced by their task environment, agencies operating in zones of conflict have come up with proposals and innovations to better recouple reliable services to improved functions, or to do adaptive management and high reliability management at the same time (table 3.2). A subset of the responses are particularly promising and these are discussed below. Together they form an emerging management regime that we can build on when trying to meet the twofold management goal.

Trading off Scale and Experimental Design in Adaptive Management

We asked our interviewees where the adaptive management components were in their programs and how these mediate the conflicting demands placed on them by the often high reliability context in which they function. Almost without exception the reply was along the lines of "Well, we are not *really* doing adaptive management, but …" This was repeated so consistently that it raises the question of whether empirically there is such a thing as the "adaptive management regime" discussed in the preceding chapter. In Florida, interviewees involved in the Comprehensive Everglades Restoration Plan insisted that they did not want to use the term adaptive management for their plan, because they felt the authors who defined the concept, C.S. Holling, Carl Walters, and Lance Gunderson, would not agree with that use of the term. And so "the Comprehensive Plan … is based on the concept of adaptive assessment" (United States Army Corps of Engineers and South Florida Water Management District 1999b, p. ii) "We are doing adaptive assessments because we haven't been able to do adaptive management in terms of large-scale experiments and controls because of political considerations," was how the lead ecologist with South Florida Water Management District explained it to us. In the CALFED Bay–Delta Program, the term adaptive management was a dominant concept in the initial documentation, but is conspicuously difficult to find in more recent documents.

What are the initiatives doing instead? If we look at the projects currently on the table or in the implementation of these initiatives, two lines of defense are offered for keeping adaptive management in zones of conflict. One is to save the large scale by sacrificing the experimental design; that is, the agencies develop large-scale plans that include an intensive "adaptive assessment" effort to monitor the proposed interventions, instead of setting up an overall experimental design. Both the Interior Columbia River Basin Ecosystem Management Project and the Comprehensive Everglades Restoration Plan are examples of this. A leading ecologist working on the former said that the plan as a whole had no experimental component to it. He hoped that some individual forest managers would be willing to set up real experiments, thereby providing a perfect example of the second line of defense, namely, the experimental design is saved by sacrificing the large scale. Here, the experiments are scaled down and the focus is on one or a few causal relationships and a limited geographical area. In the San Francisco Bay–Delta, examples of this include the Vernalis Adaptive Management Plan (VAMP) and the Battle Creek Salmon and Steelhead Restoration Project.

Figure 5.1 summarizes these tradeoffs. The axes describe scale (ranging from "small" to "large") and the extent to which the interventions are set up within an experimental design. The latter axis ranges from "low" on experimental design, that is, "normal" management interventions with limited monitoring and evaluation, to "high," where the interventions are part of a complete experimental design with control groups and controlled conditions.

The literature places adaptive management in the upper right corner of figure 5.1(a). From that theory-based starting point, the two lines of defense are to redefine adaptive management by moving it either to the upper left (more experimental design but on a smaller scale) or to the lower right (little experimental design but on a larger scale). When we look at current practice in natural resources outside the ecosystem management framework, the point of departure is the lower left corner of figure 5.1. Most natural resource management, until fairly recently, was low on experimental design and limited in scale. Our case study initiatives have pushed ecosystem management toward the theoretical ideal of adaptive management. The drive to the ideal, however, has been resisted by the countervailing high reliability requirements that have to be met at the same time. The initiatives were consequently deflected to point I (e.g., VAMP) or to point II of the lower right corner (e.g., ICBEMP) (figure 5.1(b)).

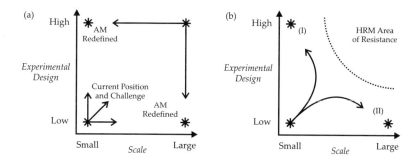

Figure 5.1 Trading-off scale and experimental design in adaptive management (AM) within a high reliability management (HRM) context. (a) Repositioning in theory; (b) repositioning in practice.

Attempts to move away from conventional natural resource policy and management towards adaptive management, while partly successful, are doomed from the start in terms of their theoretical ideal. The upper left corner of figure 5.1 is exactly where high reliability management is most uncomfortable. HROs do not want large-scale experiments with the system, because experiments mean trial and error, which test the very limits of their operations (table 3.1). Resistance to the move toward the adaptive management ideal can be visualized as the dotted perimeter line in figure 5.1(b) that demarcates an HRO's "zone of comfort." It resists experimental interventions taking place beyond that line. Giving up scale means the HRO can more easily buffer against potential errors through the redundancy in the system. Giving up experimental design allows the HRO to demand best management practices and adequate safeguards that the interventions in the large-scale plan will not jeopardize their operations.

Moves along the scale versus experimental design tradeoff are not a recoupling of adaptive and high reliability management, but rather, avoidance of that recoupling. It is a decoupling, because the line of demarcation effectively separates adaptive and high reliability management as they are actually practiced. So how do the projects in the three case studies make sense of their relation? The answer is through a storyline—basically, a variation of the much more general long-term sustainability narrative. In CALFED, the description of the "preferred alternative," the program's main proposal, states that "Improvements in ecosystem health will reduce the conflict between environmental water use and other beneficial uses, and allow more flexibility in water management decisions" (CALFED 1999a, p. ES-11). The idea is that if you do ecosystem management well, it will perforce improve high reliability management. This expectation is shared by ecologists and officials in the water agencies. An informed source within the Southern California Metropolitan Water District argues that "the better the water system performs in meeting environmental objectives, the less the regulatory screws come down on it. They might even loosen, as system health improves." In Oregon, ICBEMP's Environmental Impact Statement (EIS) program leader says, "Conceptually we argue that sustainable ecosystems will provide more sustainable services. ... The system is out of whack right now and the perturbations that we will see to get back into a sustainable situation will have an initial impact on traditional services. But it won't turn everything topsy-turvy. The system is topsy-turvy right now."

In these scenarios, the promise of recoupling is apparent, but the obstacles to fulfilling that promise are tremendous. A moment earlier, the EIS team leader noted, "We argue that large-scale experiments won't be possible. We have listed species on every acre. We are not going to be able to take short-term risks. The ESA is specifically meant to prevent short-term risks for species that are at the brink of extinction. So we are short-term risk averse. On the other hand, without taking any risks you are not going to be able to do anything. That's a crucial dilemma for our managers."

There you have it. The storyline of how ecosystem restoration will also improve ecosystem services is a promise of an end state, where any and all recoupling is by default. The story does not tell decision makers how to get to the end state. We are told, for example, that a major proposed restoration project for CALFED "should benefit CALFED's water supply reliability objective by contributing to recovery of ... Central Valley salmon and steelhead, all of which either currently, or in the future, could [positively] impact water supply reliability for water users throughout the Central Valley watershed" (CALFED, n.d.).

How is this recoupling of recovery and reliability to take place? Not only do ecologists explain that even a successful intervention will take years, if not decades, to bring key ecological functions back into a restored state; they promote adaptive management precisely because it underscores at the outset the uncertainties surrounding any claim that such an intervention could be successful. It is inconceivable that the reliability mandates, be they driven by the Endangered Species Act, water supply, or another mandate, will be put on hold for many years, while we wait for the positive effects of restoration.

Integrating Planning, Programming, and Implementation

While ecosystem restoration and adaptive management projects are repositioning themselves in zones of conflict, the focus in each case has been to make them part and parcel of comprehensive efforts to deal with the multiple issues of functions and services. The CALFED Program, the Comprehensive Everglades Restoration Plan, the Interior Columbia River Basin Ecosystem Management Project—their very names signal the commitment to connecting different program elements into one comprehensive proposal. Our discussion centers around one key example to illustrate how the CDR dynamic operates in these attempts at recoupling through comprehensive initiatives. The point is that the attempts are important, but can succeed only if they are translated into operational terms and are placed within an overall framework that enables a process of dynamic optimization among the components.

Coupling Levee Protection and Ecosystem Restoration The extended example is CALFED's Levee Protection Program (CALFED 1999e), which has been programmatically coupled with CALFED's Ecosystem Restoration Program (ERP). Levee protection is a high reliability mandate, but organized differently than water-supply management, which is a classic example of a centralized HRO with actual control rooms overseeing the entire system. Levee protection, on the other hand, is a more decentralized form of high reliability, with "control rooms" scattered throughout the Delta. Levee protection and flood management are two sides of the same coin. Maintaining and improving the levee system aim to prevent flood damage, while flood management sets out to deal with the contingencies that arise when the levee system is tested to its limits during floods. Taken together, they comprise a flood control system to prevent loss of life and property due to levee failure.

Protecting life and property is in the strong grasp of high reliability. In the San Francisco Bay–Delta, the importance of levee integrity stretches even further. There are over 100 Delta islands which, because of their below-water ground levels, exist only by virtue of the 1,100 miles of levees, play a crucial role in reliably managing water supply, water quality, and the surrounding agro-ecosystems. Levee failure in the central or western Delta would not only flood habitat and farmland (the Delta houses $500 million in agricultural production), but also disrupt or interrupt water supply deliveries to urban and agricultural users, transportation, and the regional flow of goods and services. Even if the infrastructure and facilities survived the initial effects of inundation, long-term submersion would make maintenance and repair futile. If a flooded island is not pumped dry and repaired, the resulting body of open water may well expose adjacent islands to increased wave action and additional subsurface seepage. The threat of levee failure to water quality is also readily evident and has historic counterparts. Several islands in the western Delta are vital to avert seawater intrusion into the Delta. Intrusion causes rising salinity levels, endangering water quality for urban and agricultural uses.

Against this backdrop, the Levee Protection Plan phrases its objective, "to reduce the risk to land use and associated economic activities, water supply, infrastructure, and ecosystem from catastrophic breaching of Delta levees. The goal is to provide long-term protection for multiple resources by maintaining and improving the integrity of the Delta levee system" (CALFED 1999e, p. 1-5).

Levee Maintenance and Flood Management Programs Given the reliability requirements of levee protection and flood management, to what degree does the administrative system that undertakes both display the key HRO features (chapter 4)? We first look at levee maintenance and protection, outlining the agencies involved and their mandates. At the heart of system are the approximately 50 reclamation districts, local units set up by the Bureau of Reclamation (BR). In most of the Delta, each island forms one district. A district is governed by a reclamation board, elected by landholders (mostly farmers) in the jurisdiction. Assessments are collected from the landholders to finance (part of) the district's operations. Daily management tasks are performed by a superintendent or supervisor employed by the district. Technical expertise, when needed, is provided by their district engineer, usually hired from a consulting firm.

Local reclamation districts—and in some cases other local governmental entities such as counties and special districts—are responsible for operating and maintaining most flood control levees in the Central Valley, including large numbers of privately owned levees. The districts have the responsibility, duty, and liability to maintain and operate the levees and other flood control works on a day-to-day basis in accordance with the Army Corps of

Engineers' Standard Operations and Maintenance Manual and with State Reclamation Board regulations. The local districts initiate, document, and implement projects. They can request the California Department of Water Resources (DWR) to reimburse them for up to 75% of the costs, under the Delta Levees Subvention Program (Senate Bill 34, amended by Assembly Bill 360).

The districts have the initiative in the maintenance projects. The technical expertise needed to prepare the documentation necessary for larger projects is brought in by the district engineer, occasionally assisted by biological experts and geotechnical engineers. The Department of Water Resources reviews and reimburses the projects if they comply with certain preconditions. The department also sends inspectors out into the Delta to perform "joint inspections" of some (but not all) levees, together with the local districts. When deemed necessary, the districts are asked to initiate maintenance projects to keep the levees up to standards. At the time of writing, the CALFED proposals for the subventions program (or "Base Level Protection Plan," as CALFED terms it) are to upgrade the levee safety standards to the Army Corps of Engineers PL 84-99 standard. An estimated $1 billion will be required to perform this major rehabilitation and reconstruction work on about 520 of the 1,100 miles of the Delta levees.

In addition to safety standards, projects under the subvention program are required to result in "net habitat improvement," linking levee maintenance directly to protecting and restoring Delta fish and wildlife habitat. In CALFED terms, this couples the Levee Protection Program directly to the Ecosystem Restoration Program. For the districts, such coupling operationally means that maintenance work is not only reviewed by the Department of Water Resources, but also by the California Department of Fish and Game and, with respect to the Endangered Species Act, by the Fish and Wildlife Service.

Augmenting the subvention program are the so-called Special Flood Control Projects. They cover major levee improvements or other special projects, such as combined levee and habitat restoration projects. Although responsibilities and mandates are basically the same as under the subvention program, the Department of Water Resources has a much more proactive role here. It prioritizes what types of projects (aimed at water quality, ecosystem restoration, and emergency response capability) it prefers to be implemented. Current examples are setback levees to create wetlands and riparian habitat. Most often the department initiates the special projects, in cooperation with the local districts. The reimbursement rate is usually 100%. In recent years, the Department of Water Resources has invested several million dollars in habitat construction around levees. As one official put it, "We are doing more habitat construction than anybody else in the Delta."

The third and final program of interest, alongside subventions and special projects, is flood management or "emergency response" (there are also programs on subsidence and seismic risks, but we can leave them aside). Again, the reclamation districts are the primary unit of operation. If flooding is expected, the districts set up 24-hour watches patrolling the levees. The majority (one informed insider estimates over 95%) of levee emergencies during floods are addressed by the districts themselves. The next line of defense is provided by the flood center at the Department of Water Resources offices in Sacramento. During floods, the center enters around-the-clock operations to coordinate emergency response, disseminate real-time information and keep local districts updated on weather and water-level predictions. Local agencies call on the flood center in emergencies, when they are in need of assistance such as advice or resources. The flood center dispatches staff and materials from their supply facilities in the Delta. As a last line of defense, should state and district capabilities fall short, the Army Corps of Engineers may be called in to provide disaster assistance.

High Reliability Features of Levee Protection Where are the HRO characteristics? It is clear that there is no equivalent of a Delta-wide control room such as that of the Department of Water Resources Operations Control Office for managing the State's main water supply. Still, although the line operators are decentralized and located across the Delta, reclamation districts perform many of the functions of the HRO control room. Moreover, given the nature of the system's technology, namely the levees, it is clear why this "control room" is decentralized. Reliable levee management must be localized. Failure of any one levee has immediate impacts across the Delta, which is why part of flood management is centralized in Sacramento's flood center. That said, maintenance and protection, as distinct from failure, are a highly site-specific task, which is why "95%" of the related activities are done by the local districts on their own. Contrast this with the operation of one reservoir or pumping station in the overall water supply system. Here tight coupling reigns supreme: changes in one element immediately cascade and reverberate throughout the system. This characteristic necessitates a centralized control room, whereas levee protection benefits from the more decentralized setup.

Even though or, rather, precisely because the districts are decentralized, they possess high technical competence. The line operators live literally in the system—on the Delta island—some their working lives. They have developed a long familiarity and highly specialized knowledge of the levee and related waterworks, having experienced the system over time and under different conditions. Speaking of the district reclamation boards, a water resources official summed up the point neatly, "the island is their baby." The district engineer, who has often been with the district for many years, supplements local skills and knowledge when needed (e.g., measuring slope stability, modeling deformation, doing borings). Engineers are also closely connected to the system, further strengthened by the fact that different districts share the same engineer.

The pressure for high performance and oversight comes first and foremost from the island's residents and farmers themselves, who are the front-line victims of failure. The consequences of failure reach far beyond the individual district, which is why there is a high degree of formal oversight. The clearest instances of oversight are the inspections made by the Department of Water Resources and the requirements for local districts to bring levees up to Army Corps of Engineers standards. Formal oversight has become more comprehensive over the last decade in response to levee failures in the 1980s. The failures were not catastrophic, but serious and costly enough to warrant intervention. In conjunction with expanding oversight, funding for maintenance and improvements was increased at the end of the 1980s. The events between 1973 (when SB 541 set up the subvention program) and the late 1980s establish the HRO characteristic that levee reliability is not fungible. One official said that a lack of funding was a main reason for the failures and near-misses in that period. Only when the system experienced a period of near-catastrophes did the state recognize that the price for actual failure was too high; funding had to be increased to ensure the baseline of safety.

Notwithstanding the "small kills" during development of the levee system during this century, the overall trend has been one of more or less continuous improvement. "The levee system has never been as safe as it is today," agreed one Delta levee manager, a remark we also heard from other sources, including CALFED staff. The more recent track record supports this observation. Before 1986, "we were losing islands right and left," according to a department official. However, at the time of writing no real levee failures had occurred since, even though 1995, 1997, and 1998 were flood years. "That's how we know the program is working," he concluded. The constant search for improvement is evidenced by the policy of increasingly stringent levee safety standards, but also by the special projects program. One interviewee from the Department of Water Resources estimated that over the last decade, about half of all levee protection funds had been channeled into special projects. In these projects, department resources and expertise are brought to bear on new ways to improve levee integrity while meeting other goals such as increasing channel flood capacity, habitat restoration, and water-quality reliability. A final example of the search for improvement is the work of the flood center outside the flood season. In that period, it documents in detail how the levee system behaved under flood season conditions (i.e., the location of erosion, seepage, sliding, and overtopping, among other factors). The resulting database is used to improve subsequent flood management and levee protection.

At first sight, levee construction and management may not seem to qualify as a highly complex activity. As one official stated, "We know pretty much what we need to know about the levees." This engineering knowledge is codified in a uniform standard for levee design, including the Army Corps of Engineers PL 84-99 standard supported by CALFED. What the uniform standard obscures, however, is that the causes of levee failure are highly site-specific. "Each island is different," said one engineer. For one levee the main threat comes from overtopping; for another it is seepage through the levee's foundation or sliding of the levee's slope; for another levee it is a combination of these factors and more. The performance of a levee is determined by the dynamic interplay of wide-ranging local conditions: water flows (velocity, volume, level), geophysical conditions of the soil, the shape of the levee, its history, and the material in its foundation, among other factors. Detailed engineering analysis and modeling of conditions help to develop a case-specific design. Perhaps even more important is the local knowledge of the system. "The locals have great knowledge of the situation of the levee. You need them to build *reliable* levees," one official summarized it. For that reason, officials in the Department of Water Resources and local districts termed the Corps' standard a "cookie cutter" and argued it should be adapted on a case-by-case basis in close collaboration with local districts.

A safe levee system requires constant, flexible management. There is a clear relation between use and management here, where a prevalent form of management is how the levees are actually used on a daily basis. Levees may have roads on them that are used by farmers to get around the island. This use of the levees means a constant form of monitoring and real-time detection of small changes in the system. In this and other ways, effective management is produced through redundant, multiple channels for important problem solving. Further redundancy comes about through local maintenance that occurs outside the state programs. Then there are the subvention and special projects programs aimed at maintenance and improvement. Finally, there are levee protection interventions approved and implemented on a moment's notice during flood fighting. The redundancy in the system is actively used by the local districts to get the job done. One interviewee told us that much of their maintenance work had actually been done and financed under the emergency response conditions, where earlier efforts to do them under the other programs had failed or taken too much time.

Decoupling Levee Protection and Ecosystem Restoration Perhaps the most important HRO characteristic of the levee protection program can be seen through its relation with adaptive management and the Ecosystem Restoration Program. While CALFED's Levee Protection Program underscores the need to coordinate its plan with the Ecosystem Restoration Program, its operational proposals serve first and foremost to decouple, rather than practically recouple, the two programs. "In general, it is desirable to provide separation of the habitat from the levee cross section" (CALFED 1999e, p. 4-2). As one official explained, the presence of nature or ecological values on or near the levees creates an "eternal dilemma" between the need to maintain and improve the levee and the conservation of habitat. This creates real operational difficulties. In one district, maintenance was problematic because the district had conserved and even improved habitat on the levee. As a result, all maintenance work now qualified as "habitat" and was subject to extensive environmental review under state and federal regulations. The neighboring district had decoupled by simply removing whatever habitat had remained from the levees and covering them with rock riprap. Consequently, maintenance was no longer complicated by habitat considerations. The CALFED proposals aim to solve the tensions between the two programs by taking the "ecological values" from the levee (because that space is needed for maintenance) and placing them next to the up-to-new-standard, improved levee, either waterside (preferably) or landside. This decouples the two programs in that both now have "demarcated" territories where one program has priority over the other, whereas currently their conflicting goals vie for dominance over the same areas.

We see the same commitment to reliability when links to adaptive management are raised. The first words of one Levee Protection Program interviewee were that there was nothing experimental about any of the levee improvements being proposed through CALFED. "Not much learning expected there," he said flatly. The only area for experimentation he mentioned was the integration between levee improvement and ecosystem restoration. Importantly, apart from levee setbacks, these experiments did not in any way compromise the reliability of the levee, that is, they were not really experiments with the levee, but with the surrounding habitat. The proposed levee improvements were to be built on top of the existing and properly functioning levee. As such, it was a no-risk strategy and could only make things better.

The only real experiments with levees were the so-called "setback levees." In these cases, a completely new levee would be built behind the existing one. Once it was finished, the old levee would be removed and the area in between would be turned into a wetland. Unsurprisingly, the interviewee expressed great reluctance to build this type of levee, unless its stability is all but guaranteed from a precautionary perspective of not doing anything unless it is assured safe beforehand. The trial-and-error learning that this would entail clearly conflicts with the prevailing culture of levee reliability. As yet, the only successful proposals for setback levee experiments have been those that keep the existing levee in place for several years, while careful monitoring, analysis, and management of the setback levee takes place to ensure that it is reliable before the existing levee is removed. In other words, these are experiments in the way that HROs use them: small-scale and full-scale models, outside the realm of primary operations.

The general conclusion here is that the more officials seek a programmatic separation of such complex programs as the Ecosystem Restoration Program and the Levee Protection Program, the greater the pressure to recouple them. Any recoupling will have to respect the resisting forces that produced the decoupling. While respecting the comprehensive planning and programming efforts, the only way to achieve recoupling is to look at site-specific functions and services associated with the agro-ecosystem in question, as we do in the next chapter.

Bringing Ecosystem Functions into Real-Time Management

"The environment has gotten the short end of the stick in the past," says a leading engineer in the central and southern Florida water system. "That's why the environmental water users want more water and thus a larger piece of the pie."

Regulation remains the dominant method of compelling HROs to protect or improve ecosystem functions in the process of providing of reliable services derived from the ecosystems. A barrage of standards, prescriptions, and directives constrain HRO operations, ranging from habitat protection regulations that restrict levee maintenance, to minimal flow requirements for fish passage that limit flexibility in water management, to water temperature standards to protect aquatic species that require reservoirs to release water instead of storing it to generate electricity at peak times. Many respondents expressed dissatisfaction with the results and cost-effectiveness of regulatory interventions. A Bonneville Power Administration engineer-planner describes the current system as "overly constrained." There is no solution that allows them to meet the competing objectives of reliable services and improved ecosystem functions. "So someone has to give, has to relax the constraints, but they won't," he concludes. "Long-term planning in the projects no longer has any flexibility because of the many constraints."

Meeting regulations is something quite different from improving ecosystem functions. In the best of times, standards are crude approximations of what regulatory agencies think are conditions that benefit a species or habitat. The regulations may be detailed—in fact, they usually are—but by their nature standards do not capture the natural variability of a system. Minimal flow requirements, for example, are meant to help juvenile anadromous fish reach the ocean and are tied to the periods of the year that fish passage is likely to occur; they have to be met throughout that period, regardless of whether large numbers of fish are actually coming down the river.

The inadequacy of standards in capturing ecosystem conditions and natural variability works in two ways. First, it places increased costs on the provisions of services, or even threatens their reliability, without actually being directed to protecting ecosystem functions. In the words of a representative of the Metropolitan Water District of Southern California, "We should get measures and estimates of how many fish we save when we do intervention X, such as the closing of the cross-channel gate [during the delta smelt crisis]. Do we save 1%, 10%, or 75% of the population? [The regulatory agencies] are not thinking in terms of population effects but in terms of protective measures." He immediately qualified his statement, "in reality they are thinking about it, but they can't quantify it." A similar argument was made by a CALFED insider, who said that from the perspective of accountability for these decisions "it's inevitable that they have to get these numbers."

Second, standards may fail to protect ecosystem functions in crucially important situations. Perhaps the regulatory agencies did not foresee the need to do so or protection was made impossible by meeting standards earlier. One response to this lack of foresight is a reinforced call for the precautionary principle to be applied in even more situations and for the most protective measure always to be demanded. This is already the case, according to some HRO stakeholders:

> Currently the way the pie splits up is that the regulatory agencies regulate the hell out of the water community, and then come to the table to negotiate over what is left, that is, trying to get some restoration off the ground. Their narrative for doing this is and sticking to it is their application of the precautionary principle. That means it is not enough to show that recovery may be taking place, you must show your actions do not harm any chance of recovery whether it is taking place or not. You must show every action is the most protective one. They are fed in this belief because the Endangered Species Act list isn't getting short, but longer.

Unfortunately, the precautionary principle gives no guidance whatsoever to the decision maker as to how to recover species, protect habitat, or restore ecosystem functions, let alone recouple them better to reliable services. The reason is that protective measures at one moment or location influence or even preclude measures elsewhere or later in time. An example is the regulated release of stored water early in the year to protect fish, when it is unclear whether it is really needed or whether the worst is yet to come. As one lead biologist told us, the sampling problems are insolvable, so how do you know the population is really coming down the river? Using stored water then diminishes the capacity to protect the fish or other ecosystem functions later. There are many such tradeoffs and the inflexible nature of regulation renders them difficult, if not impossible to make (chapter 3).

In short, prescriptive standards can be rigid and inefficient; they can be too broad and, thus, an inefficient use of ecosystem resources, or they can be too narrow, in which case they do not adequately protect the resources that need protection. The inflexibility of the regulatory approach precludes tradeoffs even among different ecological objectives. Most standards are species-specific, and only recently are regulatory agencies developing multispecies plans. The complications are evident. There is no regulatory answer for how to trade off different listed or near-listed species, let alone bring in a whole-system view. In an attempt to overcome the drawbacks of a regulatory approach, one noteworthy mechanism has been developed in our three case studies, the CALFED Environmental Water Account.

The Environmental Water Account The importance of the environmental Water Account for recoupling ecosystem functions and services more effectively than through regulation is evident from the hopes that the CALFED Program expresses for it:

> Through the development of an Environmental Water Account, the Program intends to provide flexibility in achieving environmental benefits while reducing uncertainties associated with environmental water requirements. Flexible management of water operations could achieve

fishery and ecosystem benefits more efficiently than a fully prescriptive regulatory approach. The Program believes that operations using an EWA can achieve substantial fish recovery while providing for continuous improvement in water supply reliability and water quality. (CALFED 1999a, p. 5.1-20)

An environmental water account, it is argued, would provide the flexibility to achieve ecosystem benefits more efficiently than a regulatory approach *and* at the same time improve water reliability. How so? There are many variants for the environmental water account's design, but the general idea is to give the regulatory agencies (U.S. Fish and Wildlife Service, U.S. National Marine Fisheries Service, and the California Department of Fish and Game) controlling access to an account filled with water or assets that are fungible with water. The account could be credited through a variety of means: "a share of water supply from new or existing facilities; variation in regulatory standards that would otherwise limit exports; by purchase of water, or by borrowing storage in new or existing project facilities. EWA assets may be in the form of water stored in surface reservoirs or groundwater storage projects, export reduction credits, or options to purchase water in the future" (CALFED 1999f, 80–81).

The account would be used to respond to real-time ecological events. Instead of trying to capture all contingencies through standards, which wastes resources ("overshooting") and hampers the flexibility of line operators, the salient regulation would now be limited to providing a baseline level of ecosystem protection. With that baseline met, the environmental water account would be used to respond to natural variability more efficiently. For example, when real-time monitoring indicates that fish are unlikely to be affected, the "export/inflow ratio" (mandating a maximum ratio of water exported from the Delta to the south compared to water entering the Delta from the north) could be "flexed" to provide water for the environmental account and to improve water-supply reliability. That water could then provide additional security in more sensitive times.

In effect, the environmental water account brings ecosystem functions into the control room as parameters that can be managed in the real-time optimization process, instead of being static constraints on the optimization, thereby serving only to undermine reliability. As such, the account is a major step toward operational recoupling. The fisheries agencies enter into direct co-management responsibility over the water system, at least for the purposes of the account, which brings them into a completely new relationship with the line operators. Critical in this regard is that the water account would force the fisheries agencies to make tradeoffs among competing ecological objectives, for example, resident fish in the dam versus anadromous fish downriver, a burden that the line operators feel is now on their shoulders alone.

Gaming exercises (more below) have identified the many positive features of the EWA, including the systemwide view of the tradeoffs surrounding real-time decisions affecting system reliability and ecosystem functions. This is in distinct contrast to the dominant regulatory approach where, according to a manager of operations, there is little communication between fisheries staff and the line operators. "They [fisheries regulators] are used to dealing with ESA stuff internally, decide what they want and then tell that to the operations people. It's not a collaborative process," he says. The EWA would change this, as it has the fisheries agencies translating their concerns into operational terms.

Yet, the water account also decouples. To see why, consider two things. First, the flexibility of the account is bought at the expense of the flexibility of the overall water supply. The dedicated environmental water is to be taken out of the much-needed redundancy of the current system. It replaces one kind of flexibility with another. "We're squeezing every last drop out the existing water supply to fill the account," as one CALFED insider phrased it. Second, the water community agrees to dedicate part of the overall water supply for environmental purposes only if it receives assurances in return. The environmental water account is to be a "no surprise" guarantee for the water community: there will be no new and unpredictable demands on "their" water in response to ecological crises. Environmental agencies will have to balance their own checkbook. An additional benefit to the water community, as noted in chapter 3, is that the water account gives the environmental community an incentive to support new storage, if only for environmental uses.

These two considerations represent the classic rationale for decoupling: the environmental water account "splits up" overall water supply and buffers environmental and urban/agricultural water users against turbulence in the other's task environment (much as new storage would do—see below). The respective users are assured their "own" water will not be jeopardized by how the others are managing (or mismanaging) their water. In addition, a "no surprise" guarantee shifts the risk toward the regulatory agencies, who would bear the burden of proof to claim more water for new endangered species listings or for unexpectedly greater needs of species already listed. The predictable response of the regulatory agencies would, however, be to insist on raising the account's "biological bar," that is, enlarge the environmental account so the agencies have a large built-in buffer to deal with the unexpected. Yet the gaming exercises found that the amount of water needed by the regulatory agencies to give a "no surprise" guarantee is not available in the current system without cutting into other water users' shares.

Decoupling environmental water from other uses may be a necessary step toward recoupling. So far, however, CALFED documentation on the environmental water account only hints at the need and opportunities for recoupling. For example, water releases for fish protection can also improve water quality by pushing saltwater intrusion away from the Delta. Certainly there is little indication that the account provides a structure for recoupling. The importance of the missing considerations is underscored by the fact that both the positive and negative features of the account's rationale could easily be extended and applied to related issues, such as when water quality was also given its own account in one of the CALFED gaming exercises. The design of the environmental water account and its accounting rules require many controversial issues to be resolved. From our view, the most important problem is the lack of clarity on how the measure fits into a long-term perspective, especially given its focus on the recovery of fish species. What happens when you take money away from long-term programs, such as the Ecosystem Restoration Program, to implement the short-term environmental water account? "My biggest concern," says one of the ecologists involved, "is how you assess the success of the account. It will be difficult to quantify how it has performed now, instead of in 10 years. In reality, there is no comparison to look for a benchmark."

Nevertheless, the EWA promises to be a major advance over the current situation, and not only for California. In the Columbia River Basin, a planner told us they had considered a similar idea: "Give everyone a budget to let them buy water for flood control or for fish, etc. Give them half a billion but then at least you know what is what. Let them spend it how they want." In Florida, a well-informed observer pointed out the similarity between the EWA and their discussion about establishing minimum flows and levels, which is also to dedicate water for environmental purposes.

The EWA—or its functional equivalent—is most developed in California. However, another important development is advanced in all three cases: a changing management structure around the control room, which seeks to better address ecosystem functions as well as services.

Interagency Management of the Control Room   Line operators in the control room focus primarily on keeping the system stable within specified bandwidths. The bandwidths include requirements regarding not only water supply, but also water quality, flood control, power generation, and other activities dependent on functioning ecosystems and landscapes. Outside the control room, these bandwidths are negotiated among the agencies involved. The planning and management of operations increasingly have become an interagency process. Each control room we visited is de facto managed by a "team" consisting of not only the HRO officials who actually operate the control room but also representatives from the Army Corps of Engineers and the Bureau of Reclamation which operate parts of the system, the state and federal fisheries agencies, and other state and federal environmental and natural resource agencies such as the Environmental Protection Agency (EPA) or the National Park Service (see chapter 4 on the importance of teams in HROs).

The planning cycle usually works from yearly plans ("long-term"), to monthly ("short-term"), weekly, daily ("short-short term"), and even hourly schedules. "In the long-term plan we look at monthly time steps as the minimum," says the operations planning analyst of the Bonneville Power Administration. The most intense agency interaction takes place around the short-term decisions. Short-term planning staff work with the other agencies through their technical management team, which meets weekly or biweekly.

CALFED has a similar institution in its "OPS [Operations] management team," as has the South Florida Water Management District around the design of its regulation schedules. These teams face difficult tradeoffs, and the frustration that sometimes comes with trying to get agreement on them is obvious, as Bonneville Power Administration's principal operations planner explains:

> There are eight or nine listed species and even more near-listed, and they all have different, conflicting demands in terms of water. There is no end to the complexity to balance the opportunities. So there is a lot to do in the short term. Even though flexibility has become smaller, people keep coming in and asking us to tweak the operations. They have a special interest or special species focus, for example, on one fish. I try to explain to them the tradeoffs, but they often don't see it or refuse to see it.

Because there are few guidelines or rules of thumb to deal with the tradeoffs, numerous decisions go unresolved. "We can't get to the tradeoffs, they just state what they want," was how one participant summarized it. Unresolved issues are then pushed to a higher interagency coordination team, which acts as a dispute resolution mechanism. So many issues are pushed up that two CALFED respondents said the management team was buried in operational issues most of the time. Even at the more policy-oriented level, it is often unclear exactly who controls what. CALFED's delta smelt crisis prompted the question of who had the authority to keep the gate closed or to open it and who was liable for the subsequent water quality problems.

While these interagency management teams could become the locus for ongoing recoupling efforts, so far agencies are struggling to set up institutional structures actually able to face up to the tradeoffs and complexities that come with bringing ecosystem functions into operations planning. Even when the interagency management teams reach agreement, the line operators might be unhappy with the outcome. A Bonneville Power Administration biologist working as an "interpreter" between the management team and the control room said, "Real-time people would get a planning document they literally couldn't read, so I help make the connections about where the fish are and what they need. I talk to the scheduling people and to marketing. They ask me, for example, whether to pick up or drop off discharge for power." For this ecologist, cost and reliability are significant issues: "Marketing is important so you know you can sell the power you are generating."

Ecologists in the Control Room   The biologist-interpreter working with the line operators is an innovation that merits attention, because she is actually in the control room and not just part of the management overseeing it. Her translation is twofold. She puts planning instructions into operational terms and she helps relate ecological information to real-time decisions on water and power generation. That is the double gap that fisheries agencies face when they try to influence real-time operations: they talk in planning terms and about the condition of fish populations. Neither translates automatically or even easily into the kind of decisions that line operators make in real time, which is frustrating to both parties. The biologist in the control room provides an ongoing form of operational recoupling.

We found a similar example in Florida, where the senior manager in the South Florida Water Management District operations office had just hired an environmental scientist to work in the control room with the technicians and engineers. The scientist is to head the stormwater treatment units, which are large wetland restoration projects. Having worked on pilot projects for these units for seven years the scientist has hands-on experience, which fits in well with the typical hands-on approach that line operators have with the system. In the words of the operations room manager, the new scientist will also function as the "translator" between the line operators and the districts' planning staff, which is where the ecologists were located.

Ultimately, the manager hoped to capitalize on "windows where we can enhance both reliability of water supply and ecosystem functions, such as pumping water out of Lake Okeechobee to free up storage capacity which would, at the same time, drive out salinity in the estuary and thus improve conditions for native sawgrass." He said there were many such windows that they could capitalize on, but currently they require deviations from regulation schedules that are subject to a relatively long and bureaucratic Corps of Engineers approval process. The process is complicated by the fact that the formal mandate for the water system is specific to flood control and water supply and does not include ecological objectives. He hoped now that the comprehensive plan was developed, they could get to a finer resolution of the proposed performance measures and indicators (which include water reliability as well as ecological standards). By giving the operations office these performance standards, it would have the mandate to make decisions locally, rather than push them up to the Army Corps of Engineers. We heard a similar remark by an operations planner in California's Department of Water Resources, who said "just give me the standards," instead of detailed planning instructions. The importance of environmental indicators for engineers is being given increasing prominence (Schulze 1999).

In short, there are important developments in bringing ecosystem functions into the control room of HROs, but they are only nascent or have had limited effectiveness so far in being recoupled with services. Not only are the necessary institutions to achieve such changes underdeveloped, but the ecosystem functions brought in are usually limited to the recovery of fish species and a short-term planning horizon at best. How these recoupling efforts relate to the long term and a whole-system perspective is unclear.

Developing Comprehensive Models and Gaming Exercises

The one place where the whole-system perspective is most concrete is in the abstract world of models. Through a patchwork of models, the ecosystem management initiatives in our three cases have explored the complex relations inside and between the natural and the high reliability systems. Nowhere have we come across comprehensive models for these regions that included both the ecosystems and the high reliability systems, although many interviewees stress the need to develop them.

Nonetheless, models developed in the three cases are impressive. In the ICBEMP, the complex of models involved over 170 data layers and cover the entire Columbia River Basin in some cases. The CRBSUM landscape model represents the most comprehensive and complex ICBEMP model. The model, introduced in chapter 3, combines current vegetation types with management alternatives in order to see how various disturbances might alter succession pathways into the future. The management interventions were assessed by their effects on ecological integrity and socioeconomic resilience. According to the ICBEMP senior scientist,

> The complex of model results and existing information was synthesized for analysis. The primary research [for such a landscape] is the synthesis, it's not an experimental design with replication. For a region as large as 145 million acres, where is the replicate? You can't do an experimental design for such a large region, the synthesis *is* the science. Information from subsampling watersheds was summarized for the statistical analysis and used in the simulations of historic and future conditions. The synthesis is, among other things, the design of new models and new measures, such as socioeconomic resiliency and ecological integrity.

The modeling is the strongest direct link of ecological science and information into we found in the case studies—a link argued to be crucial by proponents of adaptive management (chapter 3). As the interviewee made clear, it is modeling as synthesis that provides the link, not experiments as commonly recommended in active adaptive management. The models are the closest that managers and policymakers come to manipulating and experiencing surprise and learning about the ecosystem in the process. Models are able to do so because of their synthetic, generalizing character. These capabilities, in turn, most clearly distinguish modeling from much "real" experimental ecological research that all too frequently buys primary data at the price of concentrating on one or a few causal links or on small spatial plots. Such information is difficult to scale up for management implications at the landscape level (figure 5.1). The models are able to link these ideas and findings into overall system descriptions, which accounts for their core role in each of the three ecosystem management cases. "We've got to make the leap from site-specific findings to the broader landscape, with models, to make the [soil] information more practical," says one USGS ecologist (Brown 2000, p. 37).

The synthesis of ecological information is both a strength and a weakness of models. Modeling assumptions are partly supported by science, parameters are always subject to debate, and data remain incomplete and uncertain; all of which is exacerbated by the fact that the models aim to cover large, heterogeneous landscapes. Such is the decision space of ecosystem managers. Not unexpectedly, then, the interviewees involved in the modeling have frequently had to defend their work against criticism from their scientific counterparts. Or, as the leader of ICBEMP's large modeling study said, when asked whether the model was adequate, "You have to realize that models at this broad [basin-wide] scale did not exist when we started, and we developed what were the only models available for such work. Until other models come on board, they are state of the art. Deciding whether they are adequate is for later."

There have been notable modeling successes for both ecosystem management and service reliability, as noted in chapter 3, notwithstanding problems with inadequate ecological models on the ecological side. We saw that these successes are less in model prediction than in the development, comparison, evaluation, and confidence-building over different ecosystem-related alternatives and scenarios. The South Florida Water Management District lead ecologist described one important modeling exercise as follows:

> The models [used in leading up to the comprehensive plan] showed that the restoration alternatives not only helped the Everglades but also improved water reliability. Currently, [agriculture] faced water restrictions one out of every three or four years, even though there is the [governor's] mandate that they should face such restrictions not more than one year out of ten. The modeling showed that they could meet this goal. The same holds for urban users, where they face (mild) restrictions once every four or five years. This could also be reduced to once in every ten years [according to the models].

This process of confidence-building around alternatives was made possible by the patchwork of models that covered both the high reliability system as well as important ecological conditions. The natural system model and the South Florida Water Management Model helped the initiative to think through the hydrological consequences of management alternatives, including infrastructure for flood control and water supply. The predictions of changing hydrology were then used as input for the ATLSS modeling approach, which includes a high-resolution landscape hydrology model that relies on the water management model, vegetation models, spatially explicit species index models, and individual-based models of the highest trophic levels, which include animal behavior and movement.

In each of the cases, we found advanced use of sets of models covering the natural and the high reliability systems enabling confidence-building over longer term management scenarios for recoupling services and functions—an important step forward. It also has been recognized that the recoupling efforts through modeling are shot through with fundamental uncertainties and are far removed from real-time management, or even from other time horizons familiar to operations planning.

Assessments as Models Carried on by Other Means Uncertainty inherent in efforts to model complex systems have, by and large, shifted decision makers' attention from models to assessments, from adaptive management to other forms of "integrated," "comprehensive," or "adaptive" assessments, including monitoring, evaluation, and the repeated updating of original baselines and plans (chapter 3; Ayensu et al. 1999; Johnson et al. 1999). As noted previously, the Everglades Comprehensive Plan uses what it terms adaptive assessment rather than adaptive management. The goal of adaptive assessment is the continuous updating of the plan in light of new information, says the senior Everglades planner, "which drives some people crazy, who believe in here's the problem, here's the solution, so let's go onto a new problem. But you can't cookbook this stuff. [You] have to avoid having people saying, "Come back to us when you have the absolute, final answer." If you did that, you'd be doing a postmortem of the Everglades."

Through constantly improving the plan, the plan becomes the template for management in real time. "By maintaining a dynamic assessment," said the ICBEMP senior scientist, "you can have real-time impact in that any management decision bounces off the most updated assessment, and the assessment contains the best there is and all the latest models." As said earlier, in such circumstances the real science underway is the assessment's synthesis of the state of the art.

Assessments are models carried on by other means. They can be as formalized as models, and like models (e.g., ATLSS) many seek to scale up as well as scale down (see Root and Schneider 1995). The Columbia River Basin Environmental Impact Statement, according to its leader, puts into place hierarchical processes for stepping the large-scale information down to the smaller scale. The Northwest Power Planning Council (NPPC) assessment includes an expert system termed EDT (evaluation, diagnosis, and treatment). It starts at the subwatershed level and moves up in aggregating information on habitat, ecosystem functions, and species performance, incorporating a planning process that starts at the provincial level with a vision for the province and translates downward in terms of regional priorities. The Everglades plan implementation process includes a set of teams who ensure that project results fit within the overall plan's objectives just as the plan will be revised in light of these results.

As with models, assessments try to glimpse the future. "What sets our assessment apart from others," said the ICBEMP senior scientist, "is that we decided to tackle the future; that is, what would it look like if the region were managed under different scenarios?" As with models, assessments revolve around how to integrate science into the development and evaluation of such scenarios and alternatives. He continued,

> One of the things we felt strongly about was that if the [management plan] were to be sustained and defensible, it must be science-based. Yet the literature is not clear about what constitutes a science-based decision. So we had to design a process that would ensure the credibility of

the science and document its use in the decision. As part of the [science consistency] evaluation, we ask three questions: has all of the available scientific information been used? Has the decision correctly interpreted and accurately presented the science information brought forward? Have the risks associated with the decision been considered and revealed?

With respect to these risks, many interviewees—as well as the literature—call for the adoption of formalized risk assessments of different recovery and restoration options. Risk assessments promise a measure of flexibility in management that is, in the view of many, sorely lacking in the inflexible management approach of the Endangered Species Act (chapter 3).

Assessments are like models in other ways. They are not automatically relevant for management. Each of the three ecosystem management initiatives under study here has produced reports the size of metropolitan telephone books, shelves of documentation, and stacks of what at the time seemed only the necessary material. Most of these assessments are available online or on CD-ROM. But the challenge remains how to distill the sheer volume of material into on-point advice for state and national leaders, some of whom have about 10 minutes a day to devote to serious reading about such issues, according to the reckoning of one study (Katz 1993). One ecologist interviewed said she had spent days trying to figure out how one of the initiative's major planning reports had any relevance for ranchers on the ground. Nor is the problem unacknowledged by those responsible. The ICBEMP senior scientist told of producing a full report for Congress, only to learn that it had not been read, so a shorter status report was prepared. On discovering that it too had not been read, an even shorter highlights document full of graphics and fewer words was prepared. "We need a mechanism to collect, synthesize, and analyze the information so as to give decision makers a picture of what is going on before they make a decision," said the Columbia River Basin biologist planner. That has not happened yet, he added. What's the answer then? He did posit a recommendation, however: "information systems that you could hook into an MIS [management information system], which in turn would give managers a summarized picture that is relevant for them. The Web opens a lot of possibilities in this area."

A far more common answer was the need for someone to translate or interpret the pertinent results of science, research, models, and assessments, an issue noted above. "A lot of ecological information is not being used; it is lost," according to the senior Columbia River Basin biologist-planner. "We are not effectively translating it into management," he continued. One way to get better translation is to better define the roles of scientists and managers. The science consistency evaluation, in the view of the ICBEMP senior scientist, meant that "the consistency process forces dueling scientists to engage in a formal process of defending their science in the appropriate science arena instead of challenging the [resource] manager in court." A similar observation was made by a colleague, the Environmental Impact Statement team leader. Typically, she said, "management plans have multiple objectives and what is being criticized by scientists is often that managers emphasize one use over another or ignore important uses." The management decisions, she felt, are tradeoffs between resources, and the scientists are often oriented to one resource.

In the absence of formal models, clear science, and salient assessments, what is the decision maker to do? First, they and their staff are doing a great deal of formal modeling every day, though it is often not understood as such. Putting together a spreadsheet is itself the best example of the informal modeling being undertaken by more and more people today (see Wieners 1999). And spreadsheet software is as ubiquitous in ecosystem management as in high reliability operational settings, particularly the control rooms of the water-supply systems we visited.

A great deal of informal modeling goes on for both high reliability and restoration purposes. The widely commended X2 standard, which has played an important part in CALFED, is essentially an algorithm, rather than a hard prescriptive regulatory standard. It allows multiple ways to meet salinity standards related to San Francisco Bay's saltwater intrusion zone. Called "three ways to win," X2 can be measured in terms of salinity and flow conditions over different periods and at different sites. Other algorithms include the estimates, prepared by the State's Department of Water Resources, of next year's water supply based in part on a conservative "90% exceedence level," whereby water delivery estimates are generated in terms of 90% chance that the actual water supply next year will be a certain level of million acre feet based on historical figures. When asked if such estimations included formal modeling, one water resources department interviewee tapped his head and said, "Most models are up here." In describing the CALFED gaming exercise, another observer said there was a lot of "head modeling." In the Department of Water Resources calculations, line operators look at, among other things, how much water they could supply without endangering delivery in subsequent years; a range of hydrological scenarios to get a feel for the risk involved in making the deliveries; the power requirements under California's newly deregulated electricity sector; and how to meet minimum requirements for flow, instream release, water quality, and flood protection mandated by law, regulation, or agreement. Such composite analysis, estimations, and informal modeling are all directed, as an informant put it, so that the water resources department can say, "I believe I can deliver water reliability." Tacit knowledge and bounded rationality have been found to play an important role in other high reliability organizations as well (Von Meier 1999).

The Game Is the Model In zones of conflict, there is a holy grail for system modelers: a set of comprehensive models that allows operators and planners from different agencies to explore dynamic optimization processes among ecosystem services and functions in real time or over the short-term. It shines through in the demands for "measures and estimates of how many fish we save when we do intervention X, ... thinking in terms of population effects instead of what the most protective measures are" and that "it is inevitable you have to get these numbers." Responding to the current lack of adequate models or other means by which to make explicit tradeoffs and synergies between services and function, Bonneville Power Administration's operations planner said, "There is a different way to model this. You can use objective functions that optimize for competing objectives. That would be the way to go, because that would make tradeoffs explicit and allow more explicit decisions about them."

Such optimization models would enable operational recoupling. Unfortunately, they are unavailable now and this is unlikely to change in the foreseeable future. The many causal uncertainties in the links among, for instance, water flows and the recovery of fish population prohibit reliable modeling, and these links matter most to operators and planners. The difficulties in developing models that directly connect high reliability operations to the maintenance and/or restoration of ecosystem functions have, however, increased pressure to find innovative ways to learn about the "whole system" and enable dynamic optimization processes.

Yes, models are crucial here too, but in a different way than commonly supposed. Their value is not dependent on accuracy or predictive value, that is, producing the most *likely* scenarios, but on the ability to let line operators and planners learn about *possible* scenarios, what that tells them about overall system behavior, and how they might respond better to a variety of needs and events (for more on this us eof models, see note 4, chapter 3). Thus, in addition to identifying possible scenarios, the modeling allows better probing of what makes a scenario desirable or undesirable if actually realized.

One particularly successful example was "tweak week," a modeling exercise undertaken by teams assigned to generate and evaluate different alternatives from both the perspective of restoration and reliability. As described by the senior Everglades planner, there was a selection of relevant performance measures, which were then fed into an expedited alternative formulation and evaluation cycle:

> An alternative was formulated and then modeled by the modelers (about 10–12 people, mostly South Florida Water Management District modelers). The [alternative evaluation team] (30–40 interagency people) would then compare the model outcomes with the performance measures and assess the alternative. This evaluation was taken by the formulation teams to tweak and reformulate the alternative, model it, and then see whether or how it had improved. A key part of this process was the "power modeling" weekend, which remodeled the reformulated alternatives and did the model runs during the nights. ... The power modeling (initially called tweak week) built up trust, but wasn't a giant love-in ....The [team] evaluated not just ecosystem restoration scenarios, but they also had people evaluating flood and water supply performance of these scenarios. Sometimes you would have water supply [people] saying they were

happy, but ecological people saying they weren't. The process was one of constant reformulation, and they were always multipurpose, so in the end different concerns were integrated.

Features of the power-modeling weekend were also found in the CALFED gaming exercise. The games typically took place over the course of several days and involved agency officials and stakeholders, including those responsible for water supplies and listed species. They tested various scenarios under which water could be allocated month by month over the course of actual years from the historical timeline. The goal in most cases was to identify, capture, and allocate water, if necessary by finding flexibility in the water-supply system to meet both the fixed water needs in the baseline (e.g., legal commitments to meet species needs) as well as the other water supply needs above and beyond the baseline. In the view of the CALFED official responsible for coordinating the games, the gaming's chief effects were

> building confidence among participants, they trusted the gaming procedure [to such an extent that the games] got quicker plus the group who needed to be there got smaller. [The games] developed a new language, [using such terms as] backing up water, and not only at the operational level, but also at the policy level. [It helped] clarify options and made clearer what the issues were and produced an understanding of flexibility in terms of the environmental water account, which [accumulated] all the flexible water into the account but in the process took operator flexibility away to deal with real-time problems.

The process of gaming around the environmental water account helped CALFED deal with the after effects of the delta smelt crisis closing of the cross-channel gate and subsequent declines in water quality. A well-regarded EPA ecologist reported,

> The [environmental water account] thought process helped ... with the December crisis. This is a case where modeling will help us solve [a similar crisis in the future]. It will not be a problem next time. At that time, the operators [e.g., the urban water management districts] said we need the gate open all the time and the regulatory agencies said we need it to be closed all the time. The Bay–Delta Modeling Forum worked on closing the gates partially. It showed that you can do this, save more than half the fish and still achieve all the water quality standards.

Nor were CALFED and Everglades the only initiatives with success stories in modeling. The development of Bayesian belief networks through the ICBEMP exercise is also notable. Experts met together and were asked to provide their estimated conditional probabilities upon which to generate scenarios for different ranges of habitat with different ranges of outcomes for key aquatic and terrestrial species. Successful modeling also encourages its further use. As the Everglades Comprehensive Plan is implemented, the results from site-specific projects are fed back to a systemwide modeling effort to see if they confirm the original plan suppositions.

All these considerations lead to a rather remarkable conclusion: *the gaming itself is the linked model for which decision makers have been calling.* The tweak weeks, modeling workshops, and the like are the model in which decision makers must be interested if the goal is the better recoupling of services and functions. Gaming exercises are the dynamic optimization processes needed for recoupling. While current attention is focused on the seemingly intractable task of integrating and connecting the patchwork of models describing ecosystem services and functions, the game does just that, and in real time. The line operators and regulatory staff themselves function as the links that have so far escaped modeling. To put it more formally, the people provide the nonalgorithmic knowledge needed to connect the models (J. Hukkinen 2000, pers. comm.). The game is basically a simulation of system behavior, when the system is taken to include the natural and the organizational. Because the participants in the games are the links, scenarios can unfold, reverberate through the system, and produce surprises from which learning can then take place. Ecology and engineering share the notion that the chief manifestation of complexity is surprise (Demchak 1991). Learning from such surprises (i.e., learning about complexity and whole-system characteristics) is preceded by the ability to generate surprises under conditions that are not fatal or prohibitively costly to ecologists or high reliability managers. This is exactly what the gaming exercises and the related events have accomplished and brought to the ecosystem management initiatives.

The results of gaming or its equivalents were recounted to us by different interviewees in different terms, but they converge on increasing trust among participating agencies, focusing and expediting decision making by identifying the key issues; identifying the crucial gaps in modeling and research beyond the ubiquitous call for "more research"; creating a new and shared language between participants (which is then picked up by management) that expresses a better understanding of the system and its possibilities. It also inspires new policy styles, drawing the fisheries agencies out of their conventional regulatory answers and drawing HROs out of hard-infrastructure solutions, thereby proving a unique opportunity to explore the recoupling of services and functions, including generating new policy options, such as the environmental water account.

Gaming certainly appears to be a primary means to explore the recoupling of services and functions within what is basically a given water supply. For CALFED, this reflects the current political reality. The situation changes when new large-scale water storage enters the picture, as in southern Florida.

Increasing the Water Budget—The Storage Option

Storage is an important issue for all three cases, but nowhere is it more paramount than in southern Florida. "We are doubling the water budget," says a leading official of the Comprehensive Everglades Restoration Plan. "In the past, the issues have been win–lose, but now the attempt is to bring more water to the table." Water is the currency with which to buy both increased reliability and improved ecosystem functions. We return to this important point in a moment.

Why is the water budget to be doubled in Florida through storage? Water shortage proves to be largely a temporal issue. Over the whole year, Florida's rainfall exceeds the demand of all urban, agricultural, and environmental water users combined; some 50–60 inches in total. During the wet season, flood control dominates and much of the water is discharged to the ocean as fast as possible in order not to overload the water infrastructure. The discharges not only create their own ecological problems, but on a larger scale they mean water goes unused. The Everglades Comprehensive Plan estimates that on average, "1.7 billion gallons of water that once flowed through the ecosystem are wasted each day" (United States Army Corps of Engineers and South Florida Water Management District 1999a, p. 9). Only a couple of months later water is in short supply. Counties in urban areas are facing water restrictions and the Everglades National Park is drying out. Both reliability and ecosystem functions suffer as a result.

The limits to increased storage within the current system are set as much by ecological objectives as by those of water management. Lake Okeechobee, the major storage area, gradually fills up during the wet season. The regulation schedule for the lake specifies the bandwidth within which the water level is to be maintained. The upper limit for the water level is set to protect the littoral zone in the lake. Once the upper bandwidth is approached, discharges are planned, taking into account forecasted precipitation and water expected to come into the lake. First they release water to the maximum capacity of the canal system that runs through the urban areas to the ocean. That capacity is inadequate, however. Other discharges are made either to the water conservation areas (WCAs) to the south, between the lake and the Everglades National Park, or to the St. Lucie and Caloosahatchee estuaries that connect the lake to the Florida Bay in the west and the Atlantic Ocean in the east. Each discharge creates its own ecological problems. Releasing large amounts of water to the two estuaries has devastating effects: virtually all life dependent on saltwater is destroyed in these systems as they are flushed out with freshwater from the lake. "The irony is that we've made freshwater a pollutant," as one observer summed it up (Enserink, 1999b, p. 180). Releasing water to the water conservation areas uses up the remaining storage capacity there and the high water levels damage the tree islands in the areas and floods the nesting endangered wood stork, among other things. Should the district want to prevent that, they could release water from the conservation areas to the national park, but the resulting flow patterns from those releases causes damage in the park, such as flooding the nests of the Cape Sable seaside sparrow, also an endangered species. These ripple effects are key features of the tightly coupled, dynamically interactive systems discussed in chapters 3 and 4.

For the South Florida Water Management District, the situation during the wet season could be termed the "three ways to ecological failure" (cf. the "three ways to win" of the X2 salinity standard in the Bay–Delta). The three operational choices they face to manage water flows are all unappealing from an ecological point of view. One option is to release less water from the lake (i.e., raising the upper bandwidth), but this would damage the littoral zone. In fact, some environmental groups and ecologists argue that the limit should be lowered for adequate protection. Forced to release water, the water district can choose between discharges into the estuaries, which will cause damage to saltwater marine life, and discharges into the water conservation areas, causing problems there or in the national park or in both.

To summarize, the lack of storage capacity means harmful water releases in the wet season, while in the dry season it means insufficient water to adequately meet all the needs of users, including the national park. In the current system, the water district has very restricted flexibility to improve either its ecological performance or its water-supply reliability. Not surprisingly then, a major component of the proposed Comprehensive Everglades Restoration Plan is increasing storage capacity. The presettlement natural system was able to deal with this natural variability of rainfall and water flows, but most of the natural storage has been lost over time. According to a biologist at the water district, Lake Okeechobee has lost three-quarters of its original capacity.

The current proposals are to build new storage capacity: new surface water reservoirs, aquifer storage and recovery sites (more than 300 wells will be built to store freshwater bubbles, as much as 1.6 billion gallons a day, 1,000 feet underground in the Floridian saltwater aquifer), and localized storage areas in urban and agricultural areas to decrease dependency on the holding capacity of the water conservation areas. In addition to water quantity, other proposals in the plan address the quality, distribution and timing of water deliveries. These include building stormwater treatment areas (35,600 acres of constructed wetlands, adding to 44,000 acres already under construction, to treat urban and agricultural runoff water before it is released into the natural system), removing barriers to sheetflow (taking out more than 240 miles of project canals and internal levees within the Everglades to reestablish the natural sheetflow of water through the Everglades), seepage management (reducing water losses from the park to the east coast), and operational changes in water delivery schedules.

The Win-Win Promise of Storage The storyline was repeated time and again during our interviews in Florida, and is compelling as far as it goes. Increased storage will turn the zero-sum game of ecosystem restoration and water-supply reliability into a win-win for all. "We are giving everybody what they want," is how a leading official phrased it. "Of the 'new' water captured by the plan, 80% will go to the environment and 20% will be used to enhance urban and agricultural water supplies" (United States Army Corps of Engineers and South Florida Water Management District 1999a, p. 16). Combine this with the accepted notion that in southern Florida "ecosystem problems center on water," as another source puts it, and it is clear why the storyline has mobilized so much support for the plan. Even measures that do not seem directly water-supply driven turn out to be connected to it. Restoring sheet flows by taking out levees, for example, means that one of the conservation areas will lose storage capacity and is therefore only feasible because of new storage planned elsewhere.

The Comprehensive Everglades Restoration Plan presents the best case for storage as a way to reconcile ecosystem restoration and water reliability. If fact, "virtually every proposal in the plan is multi-purpose, both aimed at environmental and water-supply objectives," says a leading engineer involved in the plan. But how generalizable is this case for storage to other major ecosystem management initiatives? The engineer said that without storage, he believed CALFED would never get past the "low-hanging fruit," meaning an assorted collection of sweetened projects that contain "something for everyone," and which when consumed will leave everyone much less ready for the harder tasks ahead.

While storage is crucial for all three cases, the differences between southern Florida and the other cases have to be acknowledged. One core assumption behind the comprehensive plan is that the ecosystem problems revolve around water. "The principal goal of restoration is to deliver the right amount of water, of the right quality, to the right places, and at the right time. The natural environment will respond to these [proposed] hydrologic improvements, and we will once again see a healthy Everglades ecosystem." (United States Army Corps of Engineers and South Florida Water Management District 1999a, p. 9). While water is obviously central to CALFED and also to efforts in the Columbia River Basin, the problems there are accepted as more varied and differentiated. "Give everybody the water they need" is not a very satisfying answer, when new storage has become a necessary part of the solution for many of those involved.

Another crucial difference is the presence of Endangered Species Act-listed and near-listed anadromous fish in the Bay–Delta and Columbia River Basin. Historically, most storage capacity has been provided through dams, which present barriers to migrating fish. The storage facilities constructed to date are among the primary causes of the declining populations of anadromous fish. This has made new storage such a contentious topic in California that the CALFED agencies adopted a "preferred alternative" which explicitly postpones new large-scale storage until after further studies on feasibility have been done. Meanwhile, the ecosystem restoration program and other core programs are being implemented. In the Columbia River Basin, the situation is analogous. In fact, at the time of our interviews the topic making headlines was the proposal to take out four dams on the lower Snake River to enable better fish passage. There are no current proposals to build new large-scale storage facilities there according to our interviewees.

Notwithstanding these differences, storage remains important for all three cases because of its clear recoupling features. Storage creates redundancy in the water supply system, thereby enabling the line operators to better respond to the variability of the ecosystem and across the landscape. This variability can take the form both of variable water conditions (precipitation, flows, water quality, and so on) and of unpredictable fish movements or unplanned Endangered Species Act-driven regulatory interventions in their operations. The more general theoretical point is that storage is a form of redundancy that helps the HRO to manage the dynamic interactions among water supply, water quality, power generation, and consumption on one hand and ecosystem functions on the other. When redundancy is found wanting, the operators are no longer able to keep the system within bandwidths, as demonstrated by the delta smelt crisis in the Bay–Delta and the power blackout crisis in the Columbia River Basin.

The Problem of Storage Is Its Success Does this then mean that new storage provides the system with the redundancy it needs, thereby achieving the recoupling between ecosystem functions and services, and is it in fact the crosswalk we are looking for? Certainly, the interviewees in southern Florida agree that there is very little room for "tinkering with the system" to produce substantial improvements, which is why "we are focused on a plan to retool an existing system and its structures," in the words of the lead official of the comprehensive plan. From this perspective, efforts such as the gaming exercises around the environmental water account and the delta smelt crisis in CALFED, where every last bit of flexibility is squeezed from the system to improve recoupling, look like marginal, crisis-driven muddling-through in the absence of real structural improvements. Indeed, the same lead official remarked that, as far as he could tell, CALFED still had "to come back and agree on the harder things, like [new infrastructure such as] the peripheral canal."

It is true that CALFED and the projects in the Columbia River Basin, for that matter, face very difficult decisions regarding infrastructure to accomplish their goals. It may also be true that in the absence of new large-scale storage, CALFED is condemned to muddling-through with increasingly desperate attempts to find whatever flexibility remains in the system. But here is the key to the paradox: the muddling-through might actually be the best guarantee of achieving recoupling. The reason for this is found in the results and effects of the gaming exercises.

"Giving everybody the water they want" sounds like a formula for success, but it in fact promotes decoupling at precisely the moment recoupling is most sought after. Why? Because it eliminates the pressure to reconsider what services and functions are wanted from the system. In fact, it takes the pressure away from autonomous agencies to invest time and resources in interagency cooperation. As we saw earlier, organizational autonomy and turf have a positive role to play in ecosystem management, but only within the context where the agency's mandate, priorities, and decisions are subject to oversight and can be defended, given competing claims by different agencies on scarce resources. Yet why cooperate with other agencies if you have

Table 5.2 Gaming versus Storage: The Risks of Storage as Decoupling

Gaming	Storage
Builds trust	Reduces polarization but does not necessarily build trust for interactions afterwards
Quickens and focuses decision making	Delays or postpones hard decisions on services and functions
Moves beyond "the need for more research"	Puts off difficult questions to do more research and assessment
Identifies gaps that need to be addressed in models	Diminishes the need for comprehensive modeling and denominates everything in terms of water
Creates a new language for management and policy	Promotes the old language of infrastructure solutions
Puts interagency planning and management in touch with the dynamics and tradeoffs of operations	Relegates operations to "pushing the right buttons in the control room" in the absence of crucial tradeoffs
Inspires new policy styles, drawing the fisheries agencies out of their conventional regulatory answers (EWA) and drawing the HROs out of the hard-infrastructure solutions, thereby enabling real recoupling	Promotes business as usual, so there is no real recoupling where functions and services are redefined and optimized together, such that the only reason management is not a zero-sum game is because there is excess water
Supports dynamic optimization	Support static optimization, treating the organizational environment as constraints
Allows stakeholders to address both short-term (real-time) decisions and longer term implications by playing though scenarios month by month, year by year	Allows stakeholders to think storage will solve the problem, thereby neglecting the importance of real-time operations, and losing gains and opportunities for recoupling on the way to completing storage
Keeps the issue politicized, which sustains pressure to make auxiliary, nonwater investments in ecosystem functions such as habitat restoration	Seeks to depoliticize the issue, which in all likelihood will reduce the auxiliary, nonwater investments in ecosystem functions
Creates institutional change beyond interagency cooperation	Rewards institutional inertia by buying more time

all the critical resources you need, namely water? This is classic organizational decoupling without pressure to recouple.

The issue goes further, however, when we look at the results from the gaming exercises, undertaken in the absence of new storage, and juxtapose them against the effects of storage, when done outside a structure that increases pressure for recoupling. Table 5.2 provides that comparison.

Table 5.2's critique of storage as a policy option is not a direct criticism of the way new storage is used in southern Florida's Comprehensive Plan. Though the points raised below certainly have relevance for that case, the use of storage there is embedded in a structure that provides certain guarantees for further recoupling efforts, as illustrated by the continuing modeling exercises during the implementation of the plan. To be perfectly clear, we are not suggesting that gaming exercises produce alternatives superior to storage. Storage may in fact be a crucial part of the solution in each of the three case studies. Our point is that the gaming identifies the risks of opting for storage as the central part of the solution without also putting in place a structure that promotes, increases the pressure for, or actually leads to recoupling restored functions to reliable services. Far from being a marginal search for the last remaining flexibility, the gaming exercises and related activities contain valuable lessons on what it takes to recouple effectively.

The positive potential of the gaming exercises is not so much in the games as such. Rather, their importance lies in the fact that they effectively combine an innovative way to generate surprises from which to learn about the whole system with the pressure on stakeholders to rethink that system's functions and services. This combination is what pushes the potential of these innovations beyond "improved" interagency cooperation and into the realm of institutional change for more effectively addressing the paradox. A Bonneville Power Administration operations planner touched upon this potential when he told us, "'There is no flexibility' is what they are always saying. Of course there is capacity for the future. There are tradeoffs, so there is flexibility." Unknowingly, he hinted at an important path to the recoupling of services and functions through institutional change, to which we turn in chapter 6.

Conclusion

This chapter has taken us through some important innovations developed in the management of ecosystems in zones of conflict, as exemplified through the three case studies. What we saw is an emerging case-by-case management that is being invented yet remains largely unrecorded. The next chapter will build on these findings in developing bandwidth management, a proposal that provides our recommended crosswalk between adaptive management and high reliability management and enables the recoupling of functions and services.

6 Ecosystems in Zones of Conflict

The Case for Bandwidth Management

What profession that is core to ecosystem management is described in the following passage?

> [Their professional] representation of a ... system can be typified as physical, holistic, empirical, and fuzzy; ... [treating] the system more as a whole than in terms of individual pieces; ... [expecting] uncertainty rather than deterministic outcomes ...; [taking] uncertainty or "fuzziness" ... to be inevitable and, to some degree, omnipresent; [seeing] ambiguity ... pervade the entire system, and ... [suspecting] the unsuspected at every turn.
>
> ... [T]he underlying notion [in their professional culture] is that no amount of rules and data can completely and reliably capture the actual complexity of the system ... [I]t is more important ... for [these professionals] to maintain an overview of the behavior of the whole system than to have detailed knowledge about its components.
>
> ... [They] tend to be very wary of [the pressure to intervene], primarily because it runs counter to a basic attitude of conservatism fostered by their culture: "when in doubt, don't touch anything." Their reluctance to take any action unless it is clearly necessary arises from the awareness that any operation represents a potential error, with potentially severe consequences. (Von Meier 1999, pp. 104–107)

We suspect that many readers would see ecologists (writ large again) as the professional group whose views are being described. Ecologists, as we have seen, frequently describe the ecosystem in just such terms: it responds to external disturbances, the whole system is more than the sum of its parts, it displays nondeterministic behavior, its complexity can never be fully captured and, therefore, management is extremely challenging, with managers always reluctant to intervene—at least in major ways—in ecosystems they do not know, because this potentially creates more problems than it solves.

However, ecologists are not the group being described, and here is the surprise. Though the quoted phrases are almost textbook ecology material, the professional culture discussed here is in fact that of line operators in high reliability organizations (HROs). The descriptions are taken from Von Meier's (1999) in-depth study of the distinctly different cultures of line operators and engineers in a sample of U.S.-based HROs. We attribute the unexpected commonalties between operators and ecologists to the fact that the professional culture of both is in good part a reaction to that of the engineer planners and managers in the HROs dealing with natural resources. With respect to the systems to be managed, ecologists and line operators both end up emphasizing the characteristics that are in tension with the engineering view.

This remarkable overlap of professional cultures points to opportunities for these groups to work together more effectively than they currently do in ecosystem management—opportunities nowhere mentioned in the ecosystem management literature we reviewed. Indeed, the convergence explains why we find the presence of ecologists in the control rooms of line operators an important innovation (chapter 5). It would be foolish not to capitalize on these opportunities when decision makers are striving for more effective case-by-case management.

Professional Cultures

Where the goal is to recouple functions and services in zones of conflict, that is, to do adaptive management and high reliability management at the same time, the interactions between different professions are crucial. At the heart of the coupling–decoupling–recoupling (CDR) dynamic described in chapter 5 are the tensions and synergies between just such professional cultures. The common thread throughout the innovations we encountered in the case studies is their ability to capitalize on the synergies, or at least reduce the tensions. One sees the thread in the presence of ecologists and ecological indicators in the control room. The same holds for attempts at integrated modeling, where difficult connections are made among the patchwork of models of functions and services by means of the professionals participating in gaming exercises. The reverse also holds: the lack of innovation is often blamed on the problematic, if not absence of, cooperation between ecologists, engineers, and other involved professionals. Grumbine (1997, p. 43), for example, argues, "Few professionals have been taught to view problems in multidimensional contexts. Yet a consequence of an increasingly interdependent world is problems with multiple causes that sometimes shift unpredictably."

In more general terms, redefining and recoupling functions and services depends on cooperation among the array of professions associated with each of the functions and services at stake in the ecosystem. This conclusion also confirms why such collaboration is so difficult to realize, as the functions/services dichotomy itself almost automatically pits ecologists and biologists protecting the functions against engineers and line operators involved in service delivery. The tension between these professions, for the most part, neatly—and disturbingly—coincides with the tension between the regimes of adaptive management and high reliability management.

As soon as we move to the left or the right on the gradient—of population, resource utilization, and demand for services—relations between the professions changes. Moving to the left means that functions tend to dominate services. Analogously, ecologists such as Daily et al. (1996) claim that ecology sets the "rules of the game" for other professions to work within, based on their understanding of ecosystem functions (chapter 2). Unsurprisingly, ecologists dominate the adaptive management regime. Moving to the right raises the importance of services over functions. Here we find a similar tendency among certain groups of engineers in HROs, who present their own version of "the rules of the game." They are open to the involvement of ecologists in their operations only to the extent that such involvement does not threaten service reliability as currently defined. Accordingly, engineers dominate the regime of high reliability management. In zones of conflict, we see both tendencies vying for prominence, yet both are even less appropriate for these ecosystems than they are to the left or the right on the gradient. What is needed is a proposal for collaboration among the different professions specifically involved in case-by-case management.

This chapter identifies the normal professionalism and blind spots of each of the key professional groups and opportunities for these professions to work together more innovatively in recoupling improved functions and services. Building on these opportunities, the chapter then develops a proposal for ecosystem management in zones of conflict, described in detail for the San Francisco Bay–Delta.

Normal Professionalism and Blind Spots

Normal professionalism, a concept taken from the work of Robert Chambers (1988, p. 69), is "the thinking, values, methods and behavior dominant in a profession." In ecosystem management, normal professionalism is by necessity part of the solution, but also an in-built part of the problem. In Chambers's view, "reproduced through education and training and sustained by hierarchy and rewards, [normal professionalism] tends to specialized narrowness" (Chambers 1988, p. 69). The consequences of these institutionalized forms of "narrowness" are not hard to fathom, because the systems we are dealing with

> are exceptionally complex, with physical, bio-economical and human domains, with linkages within and between these, and with many forms

> of variance over space and time. It is as though each discipline shines a searchlight at a dark object. We assume that if all relevant disciplines are represented, the whole object will be lit up. But the beams are normally narrow with specialization. Linkages between domains are not illuminated and some zones stay dark. (Chambers 1988, p. 71)

The conclusion is clear. If we are to take seriously these systems and their complexity, we need to have an understanding of professional blind spots in order to know which system linkages are currently underexposed and insufficiently understood.

That is only one side of the coin. Here's the other: without a specific focus, that is, a certain narrowness, the professions would not have the problem-solving capacities that they have. So blind spots are a productive and necessary characteristic of professions. They are not just a sign of narrow specialization or bounded rationality at work (i.e., professionals simplifying complexity in order to understand and act). Such conveys the false impression that the persons concerned are responsible for their being narrow and blind due to their own shortcomings. More fruitful is to think of professional blind spots as "well-earned areas of ignorance," as one colleague phrased it (T. R. La Porte 2000, pers. comm.). Or to put it in the terms of chapter 4, some of the professional blind spots are quite clearly the limits to learning associated with each management regime. The goal of our analysis of professional blind spots is, therefore, not to eliminate them or to point out the shortcomings of each profession. We think that understanding what the blind spots are will point us to opportunities for recoupling ecosystem services and functions in the field more effectively.

We explore normal professionalism by bringing together characteristics from the literature and case studies for each of the key professions involved. The boundaries of the professional groups discussed here do not coincide perfectly with institutional boundaries, nor with the educational background of the professional. For example, we found a fish biologist working for a water district whose professional views better fit the description of an engineer than a species-specific biologist. We profile the professional groups key to this chapter's proposal for improved case-by-case management: line operators, engineer planners/managers, species-specific regulators, ecologist planners/managers, and modelers. Our brief profiles do not claim to fully capture all complexities and contradiction. At the risk of simplification, our aim is to capture a number of perspectives that are important to the ecosystem management initiatives central to this book. We take our lead from Von Meier's (1999) work on the professional cultures of line operators and engineers in HROs.

Line Operators

Contrary to what many outsiders think, line operators in HROs do not merely mechanically implement the operating procedures of the engineer planners and managers, working within whatever constraints they are given. They have an important and specific task to fulfill, which brings with it its own professional ideas, values, and methods. Operators must keep the system working in real time: "Unlike engineering, where the object is to optimize performance, the goal in operations is to maintain the system in a state of equilibrium or homeostasis in the face of external disturbances, steering clear of calamities. An operating success is to operate without incident" (Von Meier 1999, p. 104).

The threats to system stability are manifold: external influences, clustering of events, uncertainties in real-time system status, and even innovations in system management which may produce unforeseen effects. As Perrow (1984) has shown, such complex technological systems are inherently instable because of complex interactions and tight coupling between different units and processes (chapter 4). Line operators experience this complexity and instability differently from engineers. Their immediate, first-hand contact with the system produces the cognitive representation of the system described earlier: fuzzy, holistic, nondeterministic, full of surprises. They experience the system more as a living organism—or, in the words of one operator, the "beast"—than as a machine. This representation fits the requirements of real-time management in that it allows the operators to quickly condense a vast array of information, including gaps and varying uncertainties, into an overall *gestalt* that guides immediate action (Von Meier 1999, p. 105).

Many contingencies to which operators have to respond result in considerable discretion to make real-time management decisions. Though they work within constraints and detailed regulation schedules, contingencies tend not to be fully covered by these instructions. Operations programmed to "go by the book" restrict the improvization crucial to success in rescuing seemingly hopeless situations or in capitalizing on unexpected opportunities. Von Meier (1999, p. 106) characterizes the experience-based approach of operators as intuitive, not in the sense that it is irrational, but because it is nonalgorithmic. Ultimately, the knowledge needed to operate the system well cannot be formalized, a situation exacerbated by the fact that many system parameters are not known exactly in real time. Indeed, not all parameters can be measured in real time, measurements themselves might be wrong, or they might present contradictory indications. Maintaining a situational awareness is therefore key for line operators to keep the system stable and reliable. One typical response to uncertainties and contingencies is to build in reliable confidence intervals and bandwidths within which to keep the system, but which also provide operators with the flexibility to respond to real-time events by having room to adjust within the bandwidths. In terms of the key values of operators, reliability is of greater concern than efficiency. Their preoccupation with keeping a situational awareness also means that veracity and transparency of information is valued higher than precision or detail. "If more information has the potential to create confusion, then for operators it is bad," says Von Meier (1999, p. 106).

Predictably, line operators' professional culture has direct implications for recoupling ecosystem functions and services. While engineer and ecologist planners are developing new operating procedures to better meet reliability and ecological requirements (chapter 5), operators are wary of such interventions. Many of the new procedures reduce their margin of error, because operators are now expected to control more variables. For operators, more control is not always better, as control comes with responsibility; that is, the ability to control creates its own pressures to act (Von Meier 1999, p. 107). Such pressures run counter to the basic attitude of "when in doubt, do not touch anything." Because of their appreciation of system complexity and its nondeterministic nature, any intervention aimed at greater optimization and fine-tuning is seen as inherently threatening to system stability. Whatever form these changes take, operators will respond by stressing their need for clear and consistent operating requirements in combination with sufficient flexibility for real-time management.

Engineers

The term "engineers" denotes the group of planners and managers in HROs who design, analyze, plan for, and manage the system. A leitmotiv for their work is finding innovative solutions for problems and striving to improve things. More than line operators, engineers stress efficiency. The pressure of the new goals and requirements that their systems have to accommodate, such as fish recovery or habitat restoration, is treated primarily as an optimization problem: how can we manage water supply to most efficiently meet goals of water quality, flood control, power generation, and consumption, as well as a whole array of ecological standards and performance measures? Engineers explore the many tradeoffs among these goals through a continual process of modeling, simulation, and analysis. The aim of the process is to translate requirements into efficient operations procedures and innovations in system design, most notably new technology and infrastructure.

The search for optimization and efficiency fits the engineers' cognitive representation of the system. According to Von Meier (1999, p. 103), they consider the system

> as a composite of individual pieces, since these are the units that are readily described, understood, and manipulated. The functioning of the system as a whole is understood as the result of the functioning of these individual components ... The behavior of the system is ... abstracted and described in terms of formal rules derived from the idealized component characteristics and interactions. These rules, combined with information about initial conditions, make the system predictable.

Of course, optimization implies that causality is understood. Only then is the kind of modeling possible that is core to the pursuit of efficiency gains. Engineer planners and managers are obviously aware of uncertainties and unpredictable events. But, as Von Meier (1999, p. 103) points out, "in order [for engineers] to design and build a technical system, it is essential to understand and interpret its behavior in terms of cause-and-effect relationships. ... Stochasticity is relegated to well-delimited problem areas that are approached with probabilistic analysis." She summarizes the classic engineering representation of the system as "abstract, analytic, formal, and deterministic" (Von Meier 1999, p. 103).

The drive for efficiency and optimization is not just valued in itself; in our case studies, it is as much a response to increased pressures on the system. New demands are being made of these HROs, especially when operating in zones of conflict. An obvious example is the standards set by fisheries agencies for the HROs' operations. The engineers then try to optimize these against the other standards and goals. But of course, the basis and logic behind the fisheries' standards are very different from those of the conventional goals that the HRO tries to meet. As one engineer involved in operations planning for the California Department of Water Resources (DWR) describes,

> the fisheries people ... are not able to quantify exactly what is needed. Should the temperature of reservoir X be 60 or 61 degrees? Because of their subject matter [fish, ecosystems] they cannot be more precise. The science is not there to be more specific. I understand this, but am also frustrated by it. We need the specificity to do the optimizing and to be able to plan ahead. Without it, we are forced to be "reactive" which is increasingly difficult and costly.

Mandates to protect and restore ecosystem functions force engineers to incorporate new variables into system design and to adapt system operations. In contrast to the line operators' attitude of "when in doubt, do not touch anything," engineers have to push for innovation, even under circumstances of uncertainty and complexity. The potential for tension between the two groups is palpable. Engineers may frown on operators' objections to innovations that seem based on intractable, obscure logic or even superstition. Operators, in turn, are keen to recount situations where their intuition turned out to be more accurate than the engineers' formal predictions.

Of course, at some point, treating the competing demands on the system as an optimization problem becomes untenable. In the words of the Bonneville Power Administration (BPA) engineer planner quoted in chapter 5, they face an overly constrained system where the optimization problem no longer has a solution. "Someone has to give, has to relax the constraints," he summarizes the issue. "As long as that doesn't happen, system reliability is threatened." Under such conditions, the emphasis shifts from efficiency to reliability as the overriding concern, aligning engineers again with line operators in their mutual pursuit of reliability first.

Species-Specific Regulators

A key group in setting the constraints for the engineers and line operators is the environmental regulatory agencies. Their primary task, protection and recovery of species, is much dominated by the Endangered Species Act (ESA) and similar species-protection legislation. While their professional backgrounds may vary, species-specific biologists—in our three cases, fish biologists—seem to dominate. Their task of regulating to protect and recover certain populations—or perhaps more to the point, implementing the ESA—gives this group its distinctive professional culture.

Although more difficult to characterize from our data, our observations suggest that their thinking, values, and methods are tied not so much to the agencies where they work, but to the normal professionalism that comes with undertaking their jobs. In fact, other professionals often find the values and methods that come with these regulatory tasks difficult to appreciate and at times difficult to fathom. Consider the perspective of line operators. One senior operations planner says there is little communication between the "fish people" and line operators. The culture of the fish and wildlife professionals, he feels, is "secretive." "They are used to dealing with ESA stuff internally, decide what they want, and then tell that to the operations people," he says. Along the same lines, the water district employee quoted earlier described how an array of recovery-oriented (i.e., species-specific) standards "regulate the hell out of the water community, and then after that [the regulatory agencies] come to the table to negotiate over what is left, trying to get some restoration off the ground."

These are scarcely the neutral descriptions that could be expected from an interested party from the outside looking in. But looking from the inside out (i.e., the view from the species-specific biologists themselves) does little to negate the outsiders' description. Rather, it explains why the regulatory tasks themselves give rise to their "normal problems" and "normal solutions." Connected to their tasks is a system representation similar in some respects to that of line operators, but with radically different implications. The "system" is in principle the ecosystem, or better yet, "habitat," but in practice the regulators are almost exclusively focused on specific populations that are to be protected and recovered. Listed and near-listed populations are seen as vulnerable and directly in harm's way. Anything could push them to the edge of extinction, or even over it. Moreover, many parameters regarding their condition are unknown, uncertain, or ambiguous, including such basic information as population size and geographical distribution.

Understandably, a system thought to be vulnerable, threatened, and poorly understood gives rise to a basic attitude of risk avoidance. Rephrasing the operator's adage "when in doubt, do not touch anything," the species-specific regulator's dictum is "when in doubt, take every precaution." The attitude is anti-interventionist, like that of the operators, with the crucial difference being which system is to be shielded from the proposed interventions. When protecting fish population, anti-interventionism means countering the potentially harmful interventions of others, most notably the HROs for water supply and power generation. From the viewpoint of the engineers and line operators, the regulatory agencies' activities are, of course, experienced as highly interventionist.

According to some, the admonition, "take every precaution," places far too much emphasis is on "every." In the words of the water district employee quoted earlier, "That means it is not enough to show that recovery may be taking place, but you must show your actions do not harm any chance of recovery whether it is taking place or not." Clearly, this interpretation of the precautionary principle entails risk aversion of an overriding, tangible, short-term kind. "The ESA is specifically meant to prevent short-term risks for species that are at the brink of extinction," says the Environmental Impact Statement (EIS) project manager of the Interior Columbia Basin Ecosystem Management Project (ICBEMP). All the while, more candidate species are lining up to get on the ESA list and few are being removed from it, which reinforces the species-specific regulators' view that they are not being overly protective in the execution of their task.

Certainly, species-specific biologists are not blind to the problems that arise as a result of the short-term risk aversion that their task requires. It may even prohibit habitat restoration and rehabilitation, both seen by many ecologists as the only real, long-term solution for the survival of threatened species. The ICBEMP senior scientist remarked:

> What appears to be happening now is that there are increased limits on the activities that can take place on virtually all land. Buffer widths around riparian areas may mean that you can't go in and tinker with anything. For example, direction may state that a bulldozer could never cross a stream in any area, as it might cross where there is spawning gravel or it might disturb sediment. You have to ask how much risk in a stream reach you are willing to bear. If the answer is zero, then you have a simple answer that leads to a lot of problems. The regulatory agencies cannot define risks in a quantitative sense; they focus on describing the activities they want stopped. They cannot project the outcomes that result from actions or inactions, but they appear to know which inputs they won't allow; consequently they are managing the mix of inputs on the assumption that the resulting outcomes will be adequate. If you never allow a bulldozer in a stream, not even in a bedrock stream reach where it could pass without any real damage, then you can't take out roads, or repair a blown-out culvert. You might even have to take in equipment by helicopter, which is 10 times the average cost. To regulatory agencies [under the ESA], cost is not an issue.

All this can be traced back to the fact that the success criterion for the tasks of species-specific regulators is patently clear: to prevent the decline and possible extinction of certain species—at whatever cost, most would add. Yes, habitat is important and the preceding chapters have remarked on the drive, also among these professionals, to expand the "normal problem" to include more of the ecosystem than just the well-being of one or a few specific species. As we already noted, there is increasing interest in multispecies approaches and habitat improvement, but if species disappear, no amount of restored habitat is likely to turn that event into a success for regulatory professionals. Their accountability is framed foremost in terms of the well-being of listed and near-listed species. As many professionals are painfully aware, this focus often runs counter to a whole-ecosystem view and thus against the thrust of many ecosystem restoration initiatives.

Ecologists

A whole-ecosystem view is one of the most distinctive characteristics of the professional culture of the ecologist planners and managers involved in the ecosystem management initiatives we studied. While the institutional position of the ecologists vary, their professional culture is organized around the tasks of ecosystem preservation, restoration, and rehabilitation. The tasks put them in a crucially different position than the species-specific biologists and regulators, even though many of them share the same educational background.

Key characteristics of the ecologists were identified and discussed earlier. In their cognitive representation of the system, they share with the regulators an emphasis on complexity and uncertainty. When discussing the work done in southern Florida to establish "minimum flows and levels" of water needed to protect ecosystem functions, a well-known ecologist says, "you can't define [the minimum amount]. Adaptive management will help us learn about this, but it will never give us a number, which drives engineers crazy." Similarities between the system representations of ecologists and line operators has been stressed: holistic, fuzzy, nondeterministic, conceptualizing the system more as a whole than in terms of individual pieces, and so on. The whole-system view is especially significant because it is the product of a function-centered approach fundamental to the ecologists' thinking. Looking beyond specific species and the conditions in which they live, the ecologists' view of the system typically encompasses an array of ecosystem functions, including processes and arrangements that interact to sustain the ecosystem as a whole (chapter 2).

The function-oriented view also means that, in contrast to emphasizing the highly consequential vulnerability of specific populations, the ecosystem as a whole can be seen as relatively more resilient and able to respond to a wider range of disturbances. Resilience is a key value, more weighty than the well-being of particular species—although the latter is important of course. One species may disappear from the system, but its role in the ecosystem is sometimes taken over by another species from the same functional group (keystone species with large "umbrella" effects are the obvious exception). The value of resilience is reinforced by the argument that ecosystems can move between multi-stable states, where even though the system might make an irrevocable transition to another state and functional structure, it still possesses integrity as an ecosystem. Ultimately, the more self-sustaining the ecosystem functions, the healthier the ecosystem would be.

The function-perspective and the value of resilience that ecosystems have in this perspective means that, as one CALFED ecologist boldly puts it, "species like the Delta Smelt won't go extinct even if we would try to wipe them out, because of the resilience of the ecosystem. They survived much more extreme conditions in the past. Don't underestimate the complexity of the system." While contentious, this statement does underwrite the fact that, in comparison to species-specific regulators, some ecologists seem less reluctant to influence or intervene in the system when focusing on improving ecosystem-wide functions. In fact, ecosystem restoration is interventionist in this regard. Several ecologists we interviewed made quite clear their professional belief that, over the long term, it is not worthwhile to tinker at the margins when it comes to improving ecosystems. The system can only be saved when functions are restored or rehabilitated, even if that means short-term risk for some species, because there simply is no other choice over the longer haul. Given this strong belief, it is scarcely helpful to be told continually that the risks attached to any intervention are sufficient reasons for doing nothing. The ICBEMP senior scientist concludes, "The big challenge I see is that we can take any team of scientists into a subbasin and find 30 reasons why we should do nothing; the greater challenge is to find the 10 or 20 reasons something should be done. This is because the land management agencies have fewer and fewer options and are actually undertaking activities on fewer and fewer sites. Thus, the rationale for action is far more challenging than a rationale for inaction."

Yet, while decidedly more interventionist than the species-specific regulators, ecologists are distrustful of management for its own sake, especially when this is taken to mean mandated, conventional, natural resource management. The ecologists' guiding presupposition is not to better control or manage "the resource," but to create conditions for the ecosystem to manage and sustain itself. The standard solutions, thus, are restoring functional habitat, minimizing stressors, protecting keystone species, and rehabilitating natural processes.

The ecologists' way of collecting and using information is closely tied to their system representation. Given the complexity of the system and hence the many uncertainties in our understanding of it, much of the information decision makers currently collect is seen as of limited use. "Drop the current real-time monitoring," a CALFED ecologist told us, "At least adapt it better for the purpose of scientific experimentation. If you see delta smelt at a certain monitoring point, that doesn't mean they will turn up at the pumps. We really don't know the fish movements." The only way to get to the crucial information about the system is through controlled experimentation—that is, learning by doing, or adaptive management. This will not give a detailed picture of the ecosystem, but it will help us understand and restore some of its basic processes. In the end, detailed understanding is not only impossible, but also unnecessary for management purposes, because only the ecosystem can really manage itself. Indeed, the ultimate intervention is the one that makes further interventions obsolete, because the system has become self-sustaining.

Modelers

The importance of modelers is that their professional culture cuts across the divisions among the four groups discussed above. They share a language that allows them to work together, even though their disciplinary backgrounds are as different as the models they build. When it comes to ecosystem services and functions, modelers seek to capture the system in models that enable managers to understand, plan, test ideas, and develop policy alternatives. For instance, in our case studies, modelers were brought together by the different agencies to develop and test alternative policy scenarios for ecosystem management.

Core values for modelers are tied to the quality of their models. Quality, in turn, is defined mostly in terms of the degree to which the model is able to give reliable and valid answers to questions about the system to be managed. Values such as accuracy, precision, and predictive power are extremely important. As policymakers and managers struggle with more and more coupled problems, the comprehensiveness of the models is also crucial. Does it incorporate the key parameters at a sufficiently large scale? Modeling fish behavior in a specific upstream reach is of limited value to managers, no matter how accurate the predictions. Thus, in the three cases we studied, major efforts were made to scale up models, either geographically, in the number of parameters, or in the levels of the system to be described (e.g., developing models that connect hydrology, habitat, and higher trophic levels).

Unfortunately, the persisting tradeoff is between the models' accuracy and validity on one side and their usefulness in terms of facilitating policy and management (chapter 3) on the other. Sometimes, the tradeoff is hard to defend, even in the scientific community where modelers also present their work and of which many consider themselves a part. Remember the ecologist in charge of the comprehensive modeling for ICBEMP (p. 156), who reacted strongly to our question on the models' adequacy? We had to realize that such broad-scale models did not exist when their ecosystem management initiative started; they developed the only models available for such work. Until other models come on board, they are the state-of-the-art. Deciding whether they are adequate is for later. We suspect that many of our interviewees would have to concur.

A "normal solution" for modelers is to deal with (sub)system properties and relations that can be modeled in a valid and reliable way. As a modeler in southern Florida explained, he would sacrifice the model's usability to preserve or increase its accuracy. The tradeoff he saw in making a model more accurate was that it would also become more complex, meaning it would be more difficult to use and less transparent, two values critical for other users. That said, his group was funded to develop a PC version of their model with a more user-friendly interface, so that it could be used by stakeholders.

Not everyone is pleased with making the model more accessible and easier to use, however. The Park Service, according to the modeler, fears the model might be too optimistic. According to their own criteria, the Everglades needs more water. The model (which suggested otherwise) is therefore a threat to them. "It could be a weapon [for the Army Corps of Engineers] to say you have enough water," the modeler says; "The model makes the ACE an amateur ecologist, undermining the Park's stated position." Another way to read the observation about a model's use by other stakeholders is to reinforce a major conclusion of chapter 5: under less antagonistic conditions, the wider use of models is what has made gaming exercises so successful. As soon as one lets go of the idea that the models' foremost function is to provide accurate and valid predictions, the models can help decision makers think through recoupling options for improved functions and services.

So far, many of the links between functions and services have escaped modeling, making models difficult to explore recoupling. The value of models in the gaming exercises, as we saw in chapter 5, is not dependent on accuracy or predictive value (i.e., producing likely scenarios), but on their capacity to generate and explore a range of systemwide possible scenarios. The latter generates surprise in ways that the former cannot. Models generate surprises under conditions that are neither fatal nor prohibitively costly, which is a crucial way to learn about complexity and the connections between services and functions. We can also revert to the terms used so pejoratively above: the models allow professionals to become "amateurs" in the other professions with which they have to work. The amateur status is a threat only when the output of the models is taken at face value as authoritative predictions instead of as a means to probe and develop possible scenarios for recoupling.

With these profiles in mind, table 6.1 summarizes some of the principal features of the five professional cultures in terms of their "normal problems" and "normal solutions," again building on Chambers' (1988) work. A normal problem is the typical way in which the profession defines the problem at hand, strongly guided by methods it deems appropriate and valid. Normal solutions flow almost automatically from the normal problems. They are the conventional set of prescriptions repeated time and again in the relevant professional literature. Normal problems and solutions, in turn, lead to blind spots. Yet gaps in thinking offer new opportunities for thinking and interacting differently with respect to ecosystem management.

These cultures and their blind spots help explain why, notwithstanding the important innovations discussed in chapter 5, decoupling of programs, agencies, and professions persists. Bringing professions together in multidisciplinary teams is not enough. We need a conceptual framework to relate them to each other in a way that compensates for the blind spots and capitalizes on opportunities engendered by new interactions.

We must accept, however, that cooperation between these normal professionalisms is not enough. "Normal professionals are very good at giving the right answer to the absolutely wrong questions," one leading ecologist summarizes. "The history of the Everglades is littered with [professional blind spots]," which is why he advocates that such major ecosystem management

Table 6.1 Normal Professionalism in Ecosystem Management

Professional	Normal problems	Normal solutions	Opportunities
Line operators	Keeping the system stable in real time, operating incident free	Redundancy, flexibility, clear and realistic operating requirements	Learn how a long-term ecosystem perspective can better inform real-time management, develop ecological situational awareness in the control room
Engineer planners/managers	Meeting peakload demand reliably and safely	More efficient performance, designing, and maintaining of infrastructure	Learn how ecosystem restoration, rehabilitation, and related activities can improve service reliability
Species-specific regulators	ESA-driven protection and recovery of species	Set standards, eliminate potentially harmful interventions and short-term risks	Flex the standards, stressing the habitat provisions in the ESA and multi-species planning; identify whole-system tradeoffs
Ecologist planners/managers	Restoring or rehabilitating ecosystems to a self-sustaining, more resilient state	Restore habitat, minimize stressors, protect species, rehabilitate natural processes	Learn how system reliability can work better to restore and rehabilitate ecosystems and identify new series from improved functions
Modelers	Building better models	Improve model accuracy, precision, predictions, make more comprehensive models	Turn other stakeholders into "amateur ecologist/hydrologist/ reliability specialists," enabling gaming exercises to build trust, gain efficiency in collaboration, and produce suprise to learn about whole-system complexity

initiatives should have "meshing groups" and "skunkworks," where a variety of people meet to integrate pieces that would not be integrated otherwise. These groups ask different questions than the usual set of normal professionals and are an important source of innovations.

The Common Blind Spot

Chambers (1988) described normal professions as relatively narrow beams of light. In ecosystem management, least illuminated is the linkage between functions and services: not one of the professions has the methods or even the concepts to think systematically about the twofold management goal. Each starts either from functions or from services and then treats the other professions as posing a set of constraints or as a threat to its normal solutions. In either instance, the relation is static optimization at best—trying to optimize within the given setting. Chapter 5 showed that this means the coupled problems are still treated in a decoupled way by programs, agencies, and professions.

Organizations involved in the provision of ecosystem services—the HROs, for example—have been forced to take into account a variety of ecological parameters and indicators. Those mandated with the protection of species and the preservation, restoration, or rehabilitation of functions treat the provision of services mainly as a risk to species and a detriment to ecosystems. Both are understandable responses and, to a degree, even justified. Our cases also reveal great creativity in the ways the professionals try to make the best within what they perceive to be the given boundaries of the problem they are working on. Thus, not only are their linkages poorly illuminated, but functions and services as currently defined are taken as given—the common blind spot. Nowhere is there an active process of redefinition going on. Without redefining functions and services, improved recoupling is impossible in zones of conflict (chapter 4). We cannot resolve the paradox of simultaneously rising demand for ecosystem services and a better environment without rethinking what these services and functions are or should be in the first place. This task is at the core of our proposal for bandwidth management.

Bandwidth Management: A Proposal

What set of organizational arrangements and structures could sustain an ecosystem management process able to recouple services and functions in a zone of conflict? To answer that question, we introduce the notion of bandwidths. Afterwards, we discuss the two key processes of bandwidth management: setting the bandwidths and managing within the bandwidths. The concept of bandwidth management helps us to think about how the recoupling is currently being done and how we can improve the process, capitalizing on the innovations identified in chapter 5 and addressing the professional blind spots just discussed.

The notion of bandwidths is used by operations planners to identify the parameters and limits within which they must keep the system (see also Schulman 1993a).[1] Some parameters are given literally in the form of bandwidths, such as the regulations schedule that describes the minimum and maximum water level for Lake Okeechobee at any point in time. Other parameters function as bandwidths too, even if they do not have this specific form. Fish protection or water quality standards also define limits within which operators have to keep the system. As the number and complexity of the bandwidths have increased, so too has the size of the spreadsheets that run continuously in the control rooms to schedule and coordinate line operator tasks. The spreadsheets are themselves the product of a condensation process of an even larger number of bandwidths. From a variety of sources—such as the many regulations in the HROs' task environment—requirements are imposed on operations, often independent of each other. Operations planning translates them into a smaller, more coordinated set of bandwidths, made tractable to the operators through the spreadsheets.

All these limits define the solution space for those involved in operations: the line operators in the HROs, the engineers and regulators responsible for operations planning, but also the ecologists implementing ecosystem improvements. The bandwidths are this complex set of limits. In other words, they are the specified functions the system is to protect and the services it is to provide. The word "specified" is important, because not all functions—and to a lesser degree, not all services—can be specified in a way that is tractable for line operators. We return to this point later.

Our proposal starts from the bandwidths, because they are the focal point of many, if not most, interactions among the professionals working on ecosystem management in zones of conflict. It is here, in a variety of interorganizational processes around the control rooms, in which the bandwidths are negotiated. In other words, it is here that the conflict between different intraprogram tradeoffs and priorities is most tangible. As we see below, this tangible conflict is the main drive behind recoupling. Management of these bandwidths is, therefore, at the core of the goal to recouple improved functions and services. Bandwidths are an importantly different way to understand how tradeoffs are struck and priorities are set, both within programs and among programs.

Bandwidth management distinguishes two distinct but related processes: setting the bandwidths and managing within them. Even though both processes involve recoupling improved services and functions through dynamic optimization, distinguishing the processes is important for several reasons, both conceptual and practical. First, it is congruent with the management processes and organizational structures already in place and which we confirmed in our case studies. Also, our chapter 4 framework underscores the fact that HROs already have separated to a large extent the two processes (namely, managing within limits without testing them and setting the limits). This is reinforced by the different professional cultures of engineers and line operators. More important, the distinction helps us to make explicit the core elements of the two management regimes we are trying to reconcile. We know from both observation and theory that the control room and its real-time learning are at the core of high reliability management. We also know that adaptive management's focus is on new (i.e., experimental) ways to learn about the system over the longer term. By distinguishing the two levels of bandwidth management, we have an entry point for understanding case-by-case management in zones of conflict.

We turn now to show how the tension between the two processes provides not only challenges but also opportunities to improve ecosystem management. The role of bandwidth management is to harness the tensions in order to better recouple functions and services. As we will see, how the recoupling of improved functions and services proceeds organizationally and operationally varies in terms of whether the programs, agencies, and professionals are involved in setting bandwidths or managing within them.

Managing Within the Bandwidths

Our analysis identified innovations that operationally recouple functions and services in and around the control room. Not only were ecologists brought into the control room to work with line operators as translators of ecological information and standards. Not only did the gaming exercises enable line operators and regulators (mostly species-specific biologists) to explore new ideas from a whole-system perspective. We also saw, for example, local reclamation districts (acting as decentralized control rooms) working with ecologists to restore habitat at the same time as levees were improved. Sometimes decentralization is lateral, as in the use of a shared digital blackboard providing real-time information and decision rules for emergency response managers distributed throughout different agencies and locations (L. K. Comfort 2000, pers. comm.). We call such activities managing within the bandwidths.

Such interventions are closely tied to organizational structure. In California's DWR, for example, the new director thought it necessary to have a biologist and an engineer as his direct right hands. A senior engineer adds, "the biologists at DFG [Department of Fish and Game] and other fisheries agencies do not have a clear understanding of the water reliability (both supply and quality in the State Water Project system); the biologists at DWR do have this picture and they have been the source of more innovation, from a systems perspective, than the agency biologists." His colleague, a senior biologist at DWR, joins in by stating that "the long-term answer to the problems of getting the fisheries and water people to work together would be a cross-training program between agencies." Similar suggestions were made in the two other cases we studied. In the two others, cross-rotation was already going on, albeit on a modest scale and usually prompted more by practical concerns than by a learning strategy.

The whole-system perspective that ideally is the product of such interactions supports the dynamic optimization processes crucial to recoupling. These include the operational recoupling we find in managing within bandwidths. To reiterate, dynamic optimization means that both sides of the equation (services and functions) can be manipulated at the same time to explore tradeoffs, establish priorities, and implement the most effective management strategies from a whole-system perspective. Currently, optimizing services in relation to functions means trying to find, for example, the most cost-effective power scheduling within a set of fish protection standards or the most efficient pumping operations given salinity standards. Examples of the reverse—optimizing functions in relation to services are the process of designing the best possible adaptive management experiment, such as CALFED's Vernalis Adaptive Management Plan (VAMP), in the context of water supply and quality requirements and attempts to get as much habitat restoration off the ground as possible in the context of standardized levee improvements.

Managing within the bandwidths is about learning where the flexibility is within the current bandwidths to improve functions, while providing or enhancing reliable services. The innovations discussed in chapter 5 underscore the progress to be made here. A common theme of those innovations is that they turn what was static into a dynamic optimization process. Instead of translating the bandwidths into fixed regulatory requirements, they are now operationalized in response to real-time data, including events in functions *and* in services. In one instance, this might mean asking for more water than regulated when it is needed for functions, for example, when fish show up unexpectedly at a location. In another, water that the regulatory requirements would otherwise have claimed is freed up when the ecological conditions do not necessitate it. In this way, a minimal flow requirement may be relaxed after key fish populations have moved downstream.

Such is true operational recoupling. The bandwidths remain basically the same, but they are operationalized differently, reflecting a dynamic optimization process that enables tradeoffs to be identified and priorities set. There are concrete examples of how to achieve this. One is the CALFED Environmental Water Account (EWA), which provides a context for system-wide tradeoffs by bringing ecologists as line operators and ecological indicators into the control rooms. Another is the gaming exercises that enable operators and the fisheries agencies to "flex the standards" in real time. As a senior official in these games explained, they still realized the "template of actions" they felt were needed to protect the functions under their mandate (i.e., their bandwidths for the system) while even "generating a little more water for water users."

Bringing about dynamic optimization applies throughout the set of functions and services being managed. Although examples of the ecological constraints set for water supply are easy to find, it must be clear that the fixed constraints that services set for the restoration of ecosystem functions also have to be addressed. Take levee design. What one official refers to as a standardized "cookie cutter" was described in chapter 5 as a form of decoupling. From a dynamic optimization perspective, levee design would no longer be operationalized in a static, standardized fashion, but in tradeoff with opportunities for riparian habitat restoration. There are many ways to design levees that meet the same safety requirements (bandwidths), each with different consequences for ecosystem restoration.

Any recoupling proposal is, of course, highly dependent on the specific organizational setting in which it is to function. After all, we are talking about managing case by case. Therefore, we will discuss a proposal for CALFED and, in less detail, for the ecosystem management initiatives in southern Florida and the Columbia River Basin. Before we turn to them, however, we discuss the second core process of bandwidth management.

Setting the Bandwidths

Operational recoupling, that is, managing within the bandwidths, gets us only so far. There are obvious limits to how much flexibility for recoupling services and functions one can find within the current bandwidths. Operators continually face situations in which they cannot keep the system within all the bandwidths that are handed to them. An intervention to stay within one set leads them to transgress another. The discrepancy between conflicting bandwidths is too large to be resolved without changing the bandwidths themselves. The issue is pushed outside the control room and short-term operations planning. It moves to other organizational platforms, set up specifically to deal with what operations people generally refer to as "policy questions." To understand bandwidth management, we must start with what it means to manage within the current bandwidths. But that is only the start of the analysis. In practice, problems of operating within the bandwidths lead to considerable pressures to reassess those bandwidths by setting new or revised ones.

The platforms for reassessing and setting new bandwidths vary from individual agencies rethinking whether the current bandwidths enable them to fulfill their mandate, to interagency teams around the control rooms, such as CALFED's state-federal operations management team, to more large-scale and distributed efforts like the overall CALFED Program itself. Though the processes in these forums vary in terms of planning horizon, geographical scale, and the set of functions and services under consideration, they are tied together through the mutual analyzing, probing, and renegotiating of the current bandwidths of the system. Together, these activities comprise the process of setting the bandwidths. Sometimes this process succeeds, sometimes not. As the Environmental Impact Statement project leader said about cross-departmental collaboration, "My experience is that interagency work has been some of my best times and also some of my worst. ... If managers want to make it work they will; if they don't, it won't."

The source of the discrepancies among bandwidths is easy to locate. The overall set of bandwidths is not designed as a coherent system. Rather, it is the aggregate of an array of more or less decoupled processes and of different professional blind spots. Unlike a budget, changes in one part of the set of bandwidths can be made without automatically requiring changes in the other parts. In budgeting, the accounting system ideally does not allow for such decoupling. Changes in one entry automatically mean funds need to be transferred to or from other entries. Entries cannot be changed without explicit consequences for the overall budget. The system erects a barrier to discrepancies and protects budget integrity.

The crucial difference, of course, is that in our cases the bandwidths are not all in the same currency. Changes in one bandwidth often have complex and uncertain consequences for the others. Discrepancies may only become apparent when they produce overly constrained systems without solutions at the operational level. The closest equivalent we found for a single currency is water quantity, for which we encountered explicit budgeting metaphors. CALFED's Environmental Water Account, for example, translates regulatory and restorative actions into draws from and deposits into a water budget (chapter 5). More generally, we saw some bandwidths—and possible changes in them—denominated during operations planning by impacts on water supply. Unfeasible solutions thus surface and the individual bandwidths have to be recoupled, or at least reassessed. The EWA already has some elements of what we call "settlement templates" (more below).

Many bandwidths, however, have no direct meaningful connection to water supply—as in levee safety standards or net habitat improvement requirements. Even those that do impact water supply directly are often difficult to assess because of the many contingencies that determine actual impacts. In the Bay–Delta, the same minimal flow requirement for a specific stream might be consistent with other bandwidths in wet years or even in dry years, but might produce infeasibilities when combined with high water exports to southern California and specific gate-closing protocols in response to fish movements upstream.

In short, most bandwidths are set in a decoupled fashion. What keeps many bandwidths apart are different conceptions and operationalizations of time and scale. Competing definitions of small-scale, large-scale, real-time, and long-term are always possible. The discrepancies that arise because of their decoupled nature and the different blind spots are not just a barrier, however. They also provide opportunities to look differently at the bandwidths and to reassess how they are set. Such opportunities follow both from our positive theory of decoupling discussed in chapter 5 and from the opportunities posed by professional blind spots. In a more practical sense, the discrepancies provide opportunities to redraw the system boundaries and identify new options. When discrepancies become apparent, they must be dealt with in order to enable operations; the bandwidths are reassessed and, in practice, recoupled.

Recoupling bandwidths is an ongoing enterprise. Increasing pressure on ecosystem functions and services generates centrifugal forces which can undo an earlier recoupling. Bandwidths to protect specific ESA-listed species might be recoupled to water supply bandwidths in one year. In the next year, fisheries agencies might have to respond to still declining population numbers, and water agencies might need to respond to increased urban and agricultural demand. As each adjusts the bandwidths set for the system, new discrepancies and conflicts arise.

Resolving the discrepancies between bandwidths is a crucial process that leads us to understand how stakeholders in our case studies address the twofold management goal over time and scale. The reason, we argue, is that the decoupled bandwidths reflect the decoupling of functions and services—as the bandwidths are directly related to the set of functions and services to be managed. Functions and services are not tangible until bandwidths have been set and managed in. The recoupling of the bandwidths, therefore, is the most visible process we found in which recoupling of functions and services actually takes place—in addition to the operational recoupling achieved while managing within the bandwidth.

Identifying Tradeoffs and Setting Priorities as Recoupling

In the case studies, we recognized different ways to resolve the discrepancies between the functions and services as defined by the current bandwidths. Put in negotiation terms, there are zero-sum solutions, win-win solutions, and solutions that redefine the problem (i.e., redefine the functions and services in question).

Zero-sum arrangements are the order of the day. At all levels, from operational to policy (or long-term planning), conflicts are dealt with in ways that reduce the requirements on one or either side of functions and services. Some zero-sum decisions concern conflicting bandwidths for different functions. For example, South Florida Water Management District faces ecological problems whatever way it releases water from Lake Okeechobee (chapter 5). In the 1999 CALFED delta smelt crisis, too, either fish or water quality had to be sacrificed—or some of both.

Typical zero-sum arrangements to recouple functions and services are to insist on "relaxing the constraints" in service reliability or to heed the environmental admonition to "reduce population numbers and per capita consumption." Functions are being sacrificed for services in the one case and services for functions in the other. In each a form of recoupling is sought, in the sense that the tradeoff between functions and services is made explicit and then a new balance struck. However, while the sought-after result is a recoupled, more stable set of services and functions, the costs may be very high, as the system is pulled either to the left or to the right of the gradient.

Win-win solutions to reduce conflict between functions and services can occur in several ways. First, there are interventions that "squeeze" more from the current system without increasing extraction of the resource in the process (e.g., an urban or rural water conservation strategy). Of course, operational recoupling within the bandwidths, such as the implementation of an Environmental Water Account, is also a kind of win-win solution, but we leave operational measures aside for the moment. Another win-win solution is interventions that advance multiple objectives at the same time. These are crucially important for the recoupling effort in southern Florida, as the Comprehensive Everglades Plan's lead engineer explains: "Virtually everything in the southern Florida restudy is environmental or multi-objective. There is only one project which you could claim is single-purpose, water supply oriented and there are a few projects, most notably the taking away of barriers [to more natural water flows], that are single-purpose ecological."

In the San Francisco Bay–Delta, setback levees and the Yolo bypass were regularly cited as examples of win-win proposals. In the latter case, the restoration of a floodplain has benefited flood control and created new shallow water habitat. Not only does the habitat attract native species, it also unexpectedly provides the added functionality of being a buffer against invading exotic species: the floodplain's variable regime of wet and dry seasons generates enough variability to keep out the nonnatives. This new ecosystem function is especially valuable as many habitat restorations face not only the question, "When we build it, will they come?" but also run the risk of exacerbating problems by inadvertently promoting the further invasion of exotic species. After an unplanned experiment—a levee breach at Prospect Island—officials undertook an assessment of fish species present in the newly created habitat. Native species were not there, but the exotics were. Now the functions and services of floodplain have been redefined to include the function of buffer against exotic species, which will also benefit other functions and services in the area.

The third way to reduce conflicts is the full-blown redefinition of services and functions. As just argued, the blind spots common in the key professions involved in our ecosystem initiatives revolve around the current set of functions and services being, by and large, taken as given. Nowhere in our case studies did we find an active process of recoupling going on that redefines the functions and services of an ecosystem *in relation to each other*. We did find some evidence that redefinition is happening, but this was more a byproduct of other processes than actively pursued as such. We return to the topic of redefinition in a moment.

What is the importance of these ways to deal with the discrepancies between bandwidths? We already know that the discrepancies are the product of decoupled functions and services, which give rise to the need to set bandwidths. The processes of dealing with the conflicting bandwidths are crucially significant because these, more than anything else, force agencies to look at functions and services simultaneously, identify and explore their tradeoffs, and then set priorities in the form of new, recoupled bandwidths.

Identifying tradeoffs and then setting priorities is what has been most lacking in the ecosystem management programs so far. Yet it is the essence of the dynamic optimization processes core to any recoupling. In the words of an ecologist associated with CALFED, the program has made progress in that finally,

> people are realizing that CALFED is not a candy store. They have to come up with a plan that people are prepared to buy into. They have to be willing to pay for things, not just demand them. For years there was no sense of priorities, no swaps, no sense of negotiating, no considerations of cost effectiveness, just demands. ... Before it was to build more reservoirs, get water to all of us. Now there is a sense of high priorities, whose needs for supply and reliability are higher, ... the urban users, or the orchard growers instead of other crops. This includes water markets, but also transfers within the water export community. They are talking more about the economics of new facilities and who would pay for them. Assuming the principle of beneficiaries should pay, they are

facing up to the fact that agricultural users cannot possibly pay up for the kinds of facilities they have been demanding through CALFED.

While this interviewee focuses on the water reliability side, the same signal came from others as well, such as the Nature Conservancy and the Metropolitan Water District (MWD), who are developing "a proposal addressing the crucial problem in CALFED. ... The proposal is How to Restore the System, where 'system' captures both ecology and water structures." When we talked to the MWD representative, the proposal's content was still vague and in its very early stages. He was very specific, however, that at its core was "the need for a top-10 list of priorities that should include operations and restoration; an allocation mechanism through which CALFED distributes its funds over restoration of habitat, buying water or water options, the EWA et cetera." In this way, many of the statements quoted earlier can now be read in light of the same insistence on explicating tradeoffs: the operations planner arguing that "models [to optimize for competing objectives] have great potential to make tradeoffs explicit;" the water district interviewee demanding that "we should get measures and estimates of how many fish we save when we do intervention X"; or the CALFED official who added that for the sake of accountability, "it is inevitable that you have to get these numbers."

It might be surprising that tradeoffs and priorities were mostly lacking in comprehensive programs such as CALFED. After all, their very comprehensiveness seems the best guarantee for priorities being set (see table 5.1). But the initial, comprehensive policy couplings are notable for the absence of an answer to where the conflict is, where the tradeoffs and the priorities are. Remember that the program goals and objectives adopted by CALFED—"ecosystem restoration," "increased water supply reliability," "water quality improvements," and "enhanced Delta levee system integrity"—are to be pursued simultaneously in order to create a "win-win" resource policy. These goals and general policies are rendered into decoupled programs, and the priorities reflect this. Fundamentally lacking in the project documentation of CALFED and other initiatives is the explication of tradeoffs. Moreover, without tradeoffs among the different goals, there can be no priorities. To paraphrase Wildavsky (1979), If everything is a priority, then nothing is. We are stuck with a list of decoupled priorities which reflect intraprogram tradeoffs, which in turn compels the setting of interprogram tradeoffs and priorities.

It is important to underscore why bandwidths are used to describe this process of recoupling through striking interprogram tradeoffs and setting priorities. Why not look directly at the tradeoffs and priorities? The answer is that bandwidths are the medium through with the conflicts between intraprogram tradeoffs and priorities is articulated. Bandwidths are different from tradeoffs and priorities in three important and interrelated ways:

- Program tradeoffs and priorities are a richer set of goals, objectives and constraints than are the bandwidths that program presents to or requires from other programs. Each program has its own history, culture and record that give rise to cross-cutting, tangled goals and objectives. These are not easily operationalized in the form of bandwidths, such as meeting a range of performance standards, legal requirements or best management practices. Yet such condensation and operationalization takes place, if only for legal or strategic reasons: the program's set of complex goals can only be achieved if presented to the interagency world in the form of bandwidths with which these other agencies comply. In this way, intraprogram bandwidths are loosely coupled to intraprogram tradeoffs and priorities, albeit the degree of loose coupling varies by program and is an empirical matter.

- The pressure to generate and use intraprogram bandwidths comes about as a primary way to mediate the resource and budget conflicts among the overall programs, most notably because the control rooms need unequivocal operating requirements and limits. Interprogram bandwidths are thereby negotiated and in turn have an influence on the subsequent nature of intraprogram priorities and tradeoffs.
- Last but not least, bandwidths are necessary for high reliability organizations to achieve highly reliable operations in their control rooms (see chapter 4 on operating within limits for HROs).

Recoupling, therefore, has to start by identifying tradeoffs and setting priorities between different bandwidths and, thus, between the functions and services expressed through those bandwidths. Tradeoffs, by definition, mean recoupling; that is, exploring the relationship between two variables. The priorities that are set reflect the chosen balance between the bandwidths and ultimately between the services and functions being managed. In other words, we are talking about the policy and program equivalent of the operational dynamic optimization process. The processes of setting the bandwidths (i.e., dealing with the discrepancies and conflicts among them) seem to provide the best environment to force the organizations involved to explicate the tradeoffs and set priorities, because operations managers need unequivocal, consistent bandwidths. In more positive terms, the discrepancies point to opportunities for recoupling—again, as argued by the positive theory of decoupling and our discussion on blind spots. Of course, it remains difficult, sometimes even impossible, to nail down the exact shape the tradeoffs should take. How many fish were saved against how much water quality degradation during CALFED's delta smelt crisis? No one really knows, but at least asking the question gives some sense of balancing and trading off the risks of not knowing on either end.

To conclude, identifying tradeoffs and setting priorities regarding the bandwidths are crucial for learning. In zones of conflict the performance record will always be mixed (chapter 4). Currently, some decoupled bandwidths are frequently transgressed—or temporarily ignored—and many are transgressed some of the time. The risk is that the transgressions provide few instances for learning. Failure no longer triggers a learning event, but becomes a natural and continuous part of ongoing management. Insight into tradeoffs and priorities provides measures of failure: some transgressions are more important than others, some have higher costs than others. It is no coincidence that the precautionary principle, identified as a limit to learning in chapter 4, is famous for not allowing tradeoffs.

Rethinking the Coupling–Decoupling–Recoupling Dynamic

Our discussion of how managing within the bandwidths leads to setting the bandwidths in which tradeoffs and priorities are central, allows us to rethink the CDR dynamic. It also tells how the two core processes of bandwidth management are related.

To recapitulate, the initial, policy-level coupling of problems and issues important to ecosystem management acknowledges where the conflict is thought to be, namely, at the putative important tradeoffs and priorities. After programmatic decoupling, the discrepancies between the program's different bandwidths—between functions and services—become apparent at the operational level. There we find operational recoupling at work to reduce some of the discrepancies, if there is any recoupling to be found at all. Perhaps more important, operational recoupling makes explicit what the real conflicts are between functions and services, namely, those that cannot be dealt with at the operational level within the current bandwidths. Some discrepancies are too large to manage within the set of functions and services as currently defined. They are pushed up the organizational structures. Whatever process deals with them is constantly confronted with the tradeoffs involved; that is, it restores at least part of the initial coupling. In other words, these instances of setting the bandwidths are programmatic and policy recoupling.

Programmatic decoupling probably best captures many of the zero-sum and win-win solutions, as both establish a new balance while leaving the currently defined set of functions and services untouched (i.e., both respect programmatic, agency, and professional boundaries). However, we would also expect these conflicts to give rise to a reassessment of policy couplings, if only because some of the zero-sum decisions are politically vulnerable. Policy recoupling, then, is the redefinition of the functions and services we want from the system in light of each other and in reaction to new information and experience at the operational level. Recoupling includes redrawing the system boundaries, an activity that has profound policy implications.

From this perspective, the CDR dynamic not only demonstrates that decoupling is a necessary step in the face of turbulence and complexity, but also that it is instrumental in articulating the conflicts between functions and services—again, something that is lacking at the policy level. Operational recoupling, in turn, helps us to understand the severity of the conflict (i.e., the nature and magnitude of the discrepancies) because not all conflict can be resolved at the operational level. Conflicts that are more severe feed into processes of programmatic and policy recoupling, which are markedly different from the original "comprehensive" policy proposals. This iterative feedback is the mechanism that connects managing within the bandwidths with the setting of the bandwidths over time. It is also the process that indicates what information is best used at what scale. From this process come redefined functions and services (more below).

Redefining Functions and Services

We have discussed operational recoupling of services and functions in terms of managing within the bandwidths. Identifying tradeoffs and setting priorities for services and functions in relation to each other was discussed in terms of setting the bandwidths. In so doing, we moved from the specific to the general under current definitions of functions and services. Now we want to move to the fundamental form of setting the bandwidths: the redefinition of functions and services in relation to each other, thus opening new forms of recoupling.

The paradox of simultaneously rising demand for ecosystem services and a better environment cannot be resolved without redefining what these services and functions are or should be in the first place. While this is a crucial task, the professionals involved are least equipped for it. Our case evidence suggests that most of the energy and creativity goes into operational recoupling and into setting the bandwidths for service reliability and ecosystem improvement as currently defined. Little energy is dedicated to thinking systematically about redefining functions and services in relation to each other. In the next section we make an organizational and institutional proposal that targets this gap, but first let us take a closer look at the process itself.

In one sense, redefinition is always going on, not as a managed process, but as the de facto outcome of changing landscapes (figure 4.2). Rivers that once had floodplains are now channelized; rivers that flowed freely, allowing the passage of anadromous fish, are now harnessed as a source of hydropower; fish populations that were part of self-sustaining systems are now managed as sources of biodiversity. Such ongoing redefinition means that, except at the most general level as in table 2.1's list of functions and services, nowhere is there a given, fixed set of functions and services, the same everywhere along the gradient.

Since functions and services are not given for all cases all the time, management has to actively define them for each ecosystem. Much of this definitional work is more or less implicit, guided by what is regarded as the "natural" state for a system, an image of a specific presettlement template, or the services we already get from the system. Of course, ecologists have long put forth competing definitions and lists of ecological functions. Such definitions are not widely agreed on, nor do they look the same for different ecosystems. Redefinition is always a possibility—or actually a live probability. Since ecosystems change continuously, redefinition becomes a natural part of ecosystem management.

The most tangible evidence of redefinition at work is indeed the activities of ecosystem restoration and rehabilitation. Nothing is permanently defined or static in these activities. Any version of the presettlement template is a highly problematic goal in human-dominated ecosystems. As these problems become more and more recognized, alternatives to restoration, including rehabilitation, are proposed (chapter 2). Similarly, after criticisms by its scientific review panel, CALFED's Strategic Plan for Ecosystem Restoration (CALFED 1999d, p. 4) acknowledged that the functions that need to be restored are not a predefined given:

> Ecosystem restoration does not entail recreating any particular historical configuration of the Bay–Delta environment; rather, it means re-establishing a balance in ecosystem structure and function to meet the needs of plants, animals, and human communities while maintaining or stimulating the region's diverse and vibrant economy. The broad goal of ecosystem restoration, therefore, is to find patterns of human use and interaction with the natural environment that provide greater overall long-term benefits to society as a whole.

Of course, this does not mean the ecosystem manager is a painter working on an empty canvas with all possibilities and options open. All manner of context and path dependencies impinge on the striking of tradeoffs and setting of priorities (e.g., March and Olsen 1989). Redefining functions and services is, as we see in a moment, a highly constrained process.

While much redefinition is implicit, there are also more explicit instances. New and different functions and services are discovered during the management of ecosystems. The Yolo bypass, if you remember, took on the added function of buffering against exotic species. The Delta levees now not only serve for flood control, but also protect the agro-ecosystem functions of the Delta islands. Sometimes, the existing services are amended and added to. An EIS project manager mentioned the idea that instead of "just" logging the forest, timber extraction would provide sustainable offtake in the form of, for example, sustainable nontimber forest products, such as high-value wood products.

Some redefinitions are quite unexpected. Take the Everglades Agricultural Area (EAA). According to SFWMD's lead ecologist, "If that area didn't do what it did, if it couldn't keep the goals of the restudy, then urbanization might creep in, thereby posing an even greater threat to the natural system. Here is where agriculture is coupled to environment. [It] is the buffer against urbanization and [is] thus protecting the environment." A radical redefinition of the services and functions has been undertaken with respect to the EAA. Where its polluted runoff was once enough to warrant the area's status as a major threat to the Everglades, it is now seen as performing a necessary service to save the system—its services are recoupled to ecosystem functions in the Everglades.

A crucial push in this dynamic comes from the redrawn system boundaries, drawing them not just around the Everglades, or even around all nature areas in southern Florida, but also around the EAA and even the urban areas. This reinforces an earlier point (chapters 3 and 4), namely, a more profound way to redefine functions and services is to redraw the system boundaries. A renowned ecologist we interviewed describes the redrawing:

> In ecology, you get to draw the boundaries. Traditionally, you see a couple of different systems, for example, crops, forests, pastures. What we do is redefine the boundaries in order to see them as linked systems. In [my field sites], I'm drawing the boundaries around a set of different systems, so not around agriculture and forest et cetera, but around the whole thing ... And then you redefine the system as this whole by redefining the boundaries for the system instead of its parts. The term ecosystem plays really well into this, because you define it in a way you want. That's why I can talk about the agro-ecosystem and someone else talks about Sierra Nevada ecosystem. [The] only point is that if you call it a system, you have look at it as a system.

Now we can see that the leverage that comes from redrawing the system boundaries is not that it immediately redefines functions and services. Rather, in redrawing boundaries decision makers generate new or hitherto unrecognized options for improving ecosystem functions while at the same ensuring service reliability. Again, our interviewee gave an example of restoring a tropical forest in one of her field sites:

> To me, restoration is restoring functions. For wetlands, for example, this would be restoring filtration, potential denitrification, nutrient functions, without worrying about restoring a specific species composition. We did aim to bring back certain species, because there are only a few of them in the system, so they are important. But we also looked at restoring the degraded pastures in the neighborhood of the rain forests for certain functions, such as protecting the forest areas from fire and exotic species invasions. You would restore these pastures to act as a buffer protecting the forest from fire. In the process, you would be "restoring" something that was never there in the presettlement template. You decide about a system as a whole, and then look at functions that are necessary to meet your goal. ... You would be managing the forest for biodiversity, the pastures for productivity, and the buffers for another purpose altogether. ... The buffer itself might be managed not for integrity or productivity, but, for example, as a barrier to exotics or fire, that is its goal. Resilience might be a goal in the forest or the pastures, but not for the buffer.

There is always a need for forest fire protection, but redrawing the boundaries so that they now include the pastures as a buffer allows you to develop an ecosystem-friendly option that improves functions, while also providing the service of fire protection for the forest and feeding local farmers' livestock.

Chapter 7 discusses a similar example in a very different context: by redrawing the boundaries around the Netherlands' "Green Heart" area to include urban areas, new opportunities arise that combine services such as housing and office space with improved ecosystem functions over the landscape. These new opportunities are possible because redrawing the boundaries enables the decision maker to see both the negative and positive features of decoupling at work.

Redrawing system boundaries to redefine functions and services is more than "simple" recoupling, such as the function of an agricultural area to buffer a nature reserve against urbanization. It means actually acting with respect to functions over the whole system. For example, as we develop more fully in chapter 7, one proposal would be to undertake an ecosystem rehabilitation program across the landscape such that it includes an "urban greening" strategy to improve systemwide functions such as clean air, clean power, and clean water. The effect of thinking in terms of redrawing system boundaries can be profound. Consider calls for full-cost pricing as a way to solve environmental problems. In this narrative, the internalization of all environmental costs is the solution for improving the environment. Redrawing the system boundaries, on the other hand, is a very different way to achieve the internalization of externalities. It is not so much that costs are internalized, as the fact that urban and agricultural areas become part of the ecosystem itself. What would happen, for example, if the CALFED "problemshed" were to include the MWD of southern California as part of the system to be managed? Decision makers would immediately be required to think about the functions associated with Los Angeles for the whole system and to wonder why the MWD has no ecosystem restoration budget, say, on par with SFMWD or BPA.

The process of redefining functions and services is fundamental in setting the bandwidths—it is choosing those functions and services you want to manage for, which are then translated into bandwidths. Yet let there be no misunderstanding about the term "redefining." Again, the ecosystem is not an empty canvas offering artistic freedom for any composition. The redefinition is a highly constrained process that takes place within a preexisting biophysical and organizational context. Table 2.1 set out the resources in zones of conflict that the decision maker has to recouple—no longer the whole ecosystem, but the isolated resources: the fish, the air, the water. The resources are directly tied to services, and thus are difficult to manage and manipulate. They need to be recoupled to the wider ecosystem functions that also depend on them. Many of the services and functions already in place will necessarily be part of whatever redefinition the decision maker could come up with. Nobody in their right mind would propose drastically curtailing drinking water delivery to Los Angeles. At the other extreme, no proposal would simply "give up" endangered fish populations.

Even a relatively brief exploration such as ours comes across surprising redefinitions that strengthen the recoupling process: Levees that protect biodiversity, pastures that provide fire protection, floodplains that buffer against exotics and improve flood control reliability, gaming exercises that come up with new ways of thinking about providing water for a more broadly understood "environment." As we describe more fully below, there is great need for a program or agency mandated specifically to redefine the functions and services of the current system in light of each other. Such an agency would be able to find many more instances than the handful of examples that we, more or less coincidentally, came across. These instances would then become part of settlement templates, which brings us to the next topic of this section.

Developing Settlement Templates

Settlement templates are the guides to recoupling services and functions that decision makers have when the presettlement template is unsuitable for purposes of ecosystem management. Instead of trying to restore a system that cannot be preserved, the settlement template defines and develops the system that can be rehabilitated and maintained.

For the purposes of this chapter, each settlement template is a set of redefined functions and services within a policy structure or mechanism that brings more coherence to currently decoupled bandwidths. This definition applies to zones of conflict in which the provision of services is done by HROs under a high reliability mandate or by organizations aspiring to high reliability operations. Here, the bandwidth is the crucial interface through which much of the recoupling of services and functions is taking place. As seen in the Green Heart case study (chapter 7), there are also zones of conflict where service provision is not a real-time, high reliability activity. Housing, for example, is provided though a variety of organizations with different mandates. In such zones of conflict, the settlement template does not revolve around recoupling bandwidths as much as around other interfaces between decoupled services and functions, such as the spatial planning system in the Netherlands.

Let us take a closer look at the two principal features of settlement templates: the set of redefined functions and services, and the policy structure and mechanism for coherence. We start with the latter. The clearest example in our case studies is the Environmental Water Account. Though limited in the services and functions it encompasses (primarily for fish and water supply), the EWA is an account for water (and fungible money) governed by an accounting structure for deposits and expenditures. The accounting mechanism means that expenditures for one water use are directly tied to expenditures for other uses, thereby making visible the tradeoffs and the need to set priorities when making expenditure choices. The fundamental innovation of the accounting structure is to convert into one resource—water as a currency—the means to realize different water-use options to improve services and functions. These different options include water for VAMP, aquifer recharge, and enhanced flows in response to real-time fish passage events, among others.

Through a complex set of rules, water becomes fungible for different water-use options. Convertibility brings coherence to different water uses. For example, water stored south of the Bay–Delta in the San Luis Reservoir could be made available for uses in adjacent areas, or it could be converted into water available, say, in areas to the north of the Delta in Shasta Reservoir when priority functions there need improvement. The EWA contains water supplies of different qualities, at different locations, and at different times for converting into the different water-use options. Its accounting structure couples disparate water-use options together via convertibility, setting in place tradeoffs, forcing priority setting, and making the decoupling of the bandwidths associated with each water use more difficult.

The EWA's limitation—it decouples dedicated water for the environment from the overall water supply—would be corrected, in our view, once EWA-supported functions are recoupled with non-environmental water services, such as extra out-of-Delta water exports for the Metropolitan Water District. Currently, the EWA is much more focused on water allocations for improving functions. Indeed, the EWA is explicitly intended to buffer services, such as urban drinking water, from any turbulence associated with the functions, such as unexpected claims of the species-specific regulators on the water community. Thus, tradeoffs are made and priorities are set by the account, but these are primarily for the function side of the twofold management goal. Converting priority services into the same currency—water in the EWA—enables tradeoffs and priorities across services and functions, thereby recoupling a wider set of bandwidths and providing a barrier to their subsequent decoupling.

Let us now turn to how settlement templates redefine functions and services. Redefinition is key to resolving the paradox. It converts the paradox of seemingly mutually exclusive improvements in functions and service reliability into a relationship between improvements that is nonzero sum or even mutually advantageous. The decision maker ends up with a revised set of functions and services that has fewer zero-sum relations (or biophysical decouplings) and better enables a dynamic optimization of functions and services. By doing all this in light of the twofold management goal the decision maker sets up a barrier against subsequent decoupling.

We saw earlier that a powerful way to redefine functions and services is to redraw system boundaries. Assume the boundaries of the Bay–Delta are redrawn so that the agro-ecosystems of the Delta islands are now part of the system to be rehabilitated. Currently, the system is defined as levee protection and the maintenance of waterways within the Delta's aquatic ecosystem. As environmental stakeholders advocate taking out the levees in order to restore the aquatic ecosystem, there is presently a zero-sum relation between levee and ecosystem improvement. Redefining the ecosystem, however, from being an aquatic ecosystem to an agro-aquatic ecosystem presupposes that the levees remain. This means that the relations between levee reliability and habitat restoration would have to move out of zero sum and into joint optimization. More practically, we can ask what are the priorities for rehabilitation now. Should funds be spent on greening industrial agriculture on the island or on habitat improvement on the outside of the levee? As we saw, currently it is government that provides most of the rehabilitation funds for the levees. Redefinition from improving a levee system to improving an agro-aquatic ecosystem means that the funding structure for these improvements changes dramatically. In the United States, a generally accepted funding principle is that beneficiaries should pay. Consequently, Delta farmers and landholders would play a much more active and larger role, in which a closer relationship would exist between the services derived from the agro-aquatic ecosystem and the functions to be improved in support of those services.

Just as information is most relevant when its gatherers are actually its users, so too will farmers and landholders have the incentive to optimize both services and functions together. Imagine developing small housing settlements on an island, connected to the service of ensuring or even enhancing the quality of rural life. This service, and the increased investments that come with it, can be enhanced by building adjacent or nearby shallow water habitat (i.e., improving functions). The service of a quiet, rural quality of life may also provide the added service of keeping noisy powerboats and other recreational activities away from the rivers and streams, thereby protecting the shallow water habitat. In fact, experience suggests that local communities protecting their quality of life may provide a more effective safeguard against increasing exploitation than the fierce regulatory measures often demanded but hardly ever enacted (Fortmann 1990).

We saw how an expanded EWA provides a policy structure or mechanism that brings more coherence to the bandwidths concerning fish protection, water supply, and related functions and services. The account works within the currently defined set of functions and services, and thus its recoupling potential is limited. In the case of the redefinition of the Bay–Delta as an agro-aquatic ecosystem, the settlement template also redefines the set of functions and services in light of each other over the ecosystem as a whole. What this last example misses, however, the EWA example encompasses. The settlement template consists of *both* redefinition and structure. Since the settlement template is a key organizing device for resolving the paradox, chapter 7 turns to our final case study, on the Green Heart of the western Netherlands, to illustrate how the settlement template would work in practice.

More generally, then, the settlement template is not a master plan, nor a map, nor a blueprint. These concepts are much too static and almost always recommended in the singular—*the* master plan, *the* map, *the* blueprint. In our framework, there can never be one settlement template, but always many. Nor do settlement templates exist on their own. Rather they must be part of networks of templates, a topic we return to more fully in the next section. It is inconceivable that just one arrangement could bring together all the pertinent bandwidths and achieve recoupling across heterogeneous ecosystems and landscapes. Some templates will focus on a limited set of functions and services but over a large scale, like the EWA; others will focus on a wider set but on a smaller scale, like the Bay–Delta agro-aquatic ecosystem. Yet the small scale must be networked into the larger scale.

The process of producing the settlement templates is best conceived in terms of the gaming exercises discussed earlier that gave rise to the EWA. Remember, the gaming exercises increased trust among participating agencies, focused and expedited decision making by identifying the key issues, and identified the crucial gaps in modeling and research. They further created a new and shared language among participants (which was then picked up by management) that expressed a better understanding of the system, its tradeoffs, and priorities. Finally, the gaming inspired new policy styles, drawing the fisheries agencies away from their conventional regulatory answers and drawing the HROs out of the hard-infrastructure solutions. New policy options were thereby generated, such as the EWA. These features are precisely those on which the development and success of settlement templates rest. In our view, setting the bandwidths through settlement templates has to incorporate these features and processes, contrasting with the current process (described above in the section on "Setting the Bandwidths"), which is reactive, piecemeal, noncumulative, and without effective barriers to the persisting pressures to decouple bandwidths.

In fact, settlement templates come in all shapes and sizes, time and scale dimensions. One settlement template could bring ecologists into the control rooms of the major power and waterworks in order to improve real-time management. Another settlement template could ensure that metropolitan water districts have significant ecosystem rehabilitation and management budgets and units. A third settlement template might develop a network of ecosystem sites whose implementation of the twofold management goal synergistically connects land-use categories normally considered in competition or antithetical to each other. A fourth settlement template could combine elements of all these three. A fifth template could be markets that are able to internalize externalities as a way to correcting market failures. For example, the power behind many proposals for water markets is their settlement template nature, that is, the privatization redefines ecosystem services and provides a structure that has incentives for recoupling (in this case, through the buying and selling of water).

Whether they are in fact settlement templates depends on whether they redefine functions and services and provide a structure that is a barrier against pressures to decouple. In the first example of the preceding paragraph, the settlement template would arise because ecologists, engineers, and line operators are redefining the system from which service reliability is produced; for example, through the Bay–Delta modeling which provided a way to meet fish recovery and urban water standards in response to the delta smelt crisis. In the second example, MWD would use its environmental restoration budget to develop wetlands in its own reservoirs, as a way of improving water quality there. This would internalize some of the hard choices between rehabilitation and service reliability, thereby forcing MWD to face tradeoffs and set priorities in making those choices. In the third example, the settlement template is much like that we will describe for the Netherlands' Green Heart in chapter 7. Many examples generate processes that mimic the gaming exercises.

Many of the tradeoffs embodied in the settlement template are made all the time, but currently remain hidden behind the rhetoric of each decoupled program, agency, or professional group in their attempts to justify and be accountable for their activities. The fisheries agencies by mandate must argue that they are trying to save all the listed and near-listed species, just as by mandate the HROs must argue that they are not sacrificing service reliability. Together, they keep restating the fundamental paradox at the heart of this book. In reality, both are forced to accept tradeoffs and set priorities; that is, to make the hard but necessary selections of what is important. Under the status quo, which this book wants to change, not all the species will be saved, or saved in the same way or with the same degree of success. Neither will all services be delivered in the same way, against the same price or with the same degree of reliability.

The current decoupled bandwidths and plans obscure the tradeoffs and how they are struck, implicitly or otherwise. The settlement template draws out the implicit priorities and makes more explicit the intrinsic flexibility in the system. In other words, since not all decoupled bandwidths can be met, choices are made about which ones are most important—if not explicitly, then by nondecisions, with equally far-reaching, albeit unmanaged, consequences. Indeed, the virtue of settlement templates is that they use the conflict over ecosystems and service reliability in order to compel tradeoffs and priorities for the multiple ways a given amount of money can be spent to better address meeting the twofold management goal.

The key is to get the tradeoffs out in the open. For ecosystem functions this means the settlement template requires difficult ecological priorities, often with implications for services. But it also opens new pathways for defining a system that can be preserved. Such a process would, for example, begin by asking, what ecosystem or set of ecosystems is implied or defined when we start by protecting species 1, 12, 18, 33, and 36 from the listed and near-listed species? What new services are or can be provided by the actions to protect or otherwise recover these species and their habitats? What are the time and scale dimensions at which all of this is to operate? Such a process addresses the many tradeoffs that are now ignored, dodged, or overlooked, to the dismay of the HROs who feel they are bearing the brunt of the tradeoffs, while this is not their job in the first place. To name a few:

- *Tradeoffs between different species.* We heard numerous examples where different species in the same system were pitted against each other. For example, drawing down reservoirs to help the passage of anadromous fish could have negative impacts on the resident fish populations.

- *Tradeoffs between aquatic and terrestrial ecosystems.* The ICBEMP senior scientist described how controlled burning for fuel load management in the forests harms not only air quality conditions, but also vulnerable aquatic ecosystems. Where there once were pulses of sediment working their way through the system due to large fires, the situation now is one in which "the system is chronically bleeding sediment." The strong aquatic populations are all isolated or fragmented, so "we want to deal with uphill conditions in these areas, and the aquatic folks say don't tinker with it, it's too risky."
- *Tradeoffs between species and other ecosystem considerations.* This relates to an ongoing discussion in conservation biology. "While detaching species from management goals, some authors have attempted to justify paradigms with appeals to notions of habitat, 'quality,' or 'integrity,' and the values of rare species are thus articulated as indicators of some attribute of natural areas rather than as elemental concerns in and of themselves" (Goldstein 1999, p. 248).
- *Tradeoffs between long-term and short-term factors.* Most notable here is the ESA's purposeful avoidance of short-term risk undermining, if not forestalling, attempts to improve the long-term habitat conditions of the ecosystem.
- *Supply and demand tradeoffs.* An example here is whether to build new water storage infrastructure to increase operational flexibility or keep the pressure on to increase water conservation efforts in the absence of such storage.
- *Tradeoffs between settlement templates within a network of templates.* Since templates are always plural and operate best within a network that connects them for any given landscape (more below), the network also reveals tradeoffs and priorities between templates.

In the context of the (networks of) settlement templates, many of these tradeoffs lead to the identification of keystone species—species that are crucial to the structure and function of ecosystems and are indicative of the success of the recoupling. Keystone species provide a "crosswalk" or language that effectively translates and connects the different management regimes. Ecologists, as a professional group, are comfortable with "unmanageabilities," such as "the whole ecosystem" or "the long term." Of course, this is in part because they are wary of conventional management in general and would usually rather reduce the need for management in any form (e.g., Orr 1999). HROs' engineers and line operators, on the other hand, find it difficult to deal with unmanageabilities. They can deal better with species, however. The task of the settlement template is to make sure the keystone species are tied to the functions that are to be preserved and improved as well as to the services that have priority. They are the physical recoupling between the real-time management of the control room and long-term rehabilitation of the ecosystem—or, to put it differently, between high reliability management and adaptive management. The idea could be extended further to include keystone processes and activities that bring the long term and whole system into the control room. Keystone processes are the physical recoupling of functions and services, which, when known, HROs can manage for. Unfortunately, many current proposals for preserving and maintaining keystone species are decoupled from each other and made outside a settlement template framework

To conclude, let it be clear that facing the difficult tradeoffs inherent in the settlement template or within a network of templates does not imply a fatalistic or overly pragmatic scenario: giving up the currently threatened ecosystem functions in the face of population pressures and extraction. First, refusing to make such tradeoffs leaves us clinging to a system that cannot be preserved or restored anyway. Second, the balance between functions and services can be tilted either way, expressing the social, political, and cultural values in which the ecosystem management initiative is embedded. If improving functions is more of a priority, then that can be expressed through a settlement template or networks of them. This could even go as far as reducing population growth and per capita consumption in the region. Such efforts move the system from its zone of conflict to the left of the gradient; that is, into the realm of the adaptive management regime. Still, for many systems this will not happen. Thus there will always be the need for settlement templates that revolve around systems that *can* be saved and improved. Politics matter, but let the politics revolve around the issues that matter, namely, what settlement templates we want for the zones of conflict, in which real people live and work in real time.[2]

Networking Settlement Templates

Ecosystem management initiatives that create and revolve around settlement templates of ecosystems within a landscape raise the question of how these templates are connected to one another. Networks that connect settlement templates, in turn, beg an even wider question: how do the networks mimic or match the whole system? Even if we could assure consistent management strategies across initiatives, how would scattered restoration or rehabilitation sites improve the wider system of which they are part? It is one thing to argue that policy should ensure the simultaneous management of sites that are, say, agricultural and aquatic, as in the Bay–Delta, but it is a completely different task to show how the sites themselves add up to a "healthier" agro-aquatic ecosystem or, for that matter, to a healthier California and beyond. In answering the wider question, the networked settlement templates become clearer. The next chapter details a proposal for one such network.

The canonical answer to how sites reflect and are linked to the whole, and one to which we subscribe, is to ensure that (1) the sites chosen for improving ecosystem functions and services are representative of other sites such that what is learned at one site can be generalized to the others, and (2) the sites themselves incorporate keystone processes and species, which if improved will necessarily lead to wider improvements. Obviously this should be done where possible. Yet, as we note throughout this book, the decision space in which ecosystem management takes place is one of persistent uncertainty, complexity, and incompleteness along all the dimensions that matter—ecological, organizational, political, and societal. It is not always possible to know what is keystone, let alone the universe from which one is sampling for representative sites and cases. How then does the part fit into the whole, when the whole is complex, uncertain, and incompletely known?

The small scale is clearly problematic for each of the four management regimes. Saving patches of remaining "wilderness" is hardly acceptable to advocates of self-sustaining management. Advocates of adaptive management recommend large-scale interventions because the findings and results of small-scale experiments are so difficult to generalize and scale up over the landscape and beyond. The latter holds mores for case-by-case management, whose success, let alone "generalizability," is evident only at the small scale of locality and site. High reliability managers, for their part, see themselves as increasingly hamstrung by the small scale carried to its logical extreme; namely, saving that fish in that river, regardless of cost. More formally, how do small-scale sites and recoupling interventions add up to large-scale change across settlement templates?

The answer to these questions, we believe, requires a full-blown, positive theory of the small scale that moves beyond conventional terms of casting the small scale as the micro level in search of a macro level that establishes the former's overall meaning. A positive theory must justify the small-scale focus in terms other than statistical representativeness of small-scale findings about a species or habitat or the ease of scaling up and generalizing findings associated with any one settlement template to the whole ecosystem, landscape, or beyond. We are now in a position to provide such a theory and understand in the process how networks of settlement templates enable us to address the system as a whole.

The core of the positive theory of the small scale was initially set out in chapter 5. There, the negative and positive features of decoupling were also discussed, but in organizational, programmatic, and profession terms (table 5.1). In the process, the scale dimensions of such decoupling were submerged. Table 6.2 reveals the scale considerations, since virtually the same negative and positive features characterize that other form of decoupling: the small scale separate and distinct from the large scale.

Table 6.2 lists complaints commonly made against the small scale within a context that privileges the large scale as the better option for governance, oversight, and coordination. Yet just as decoupling can be positive for organizations, programs, and professions, so too can it be for the small scale, and in the same way. As we saw in earlier chapters, the primary response to a turbulent task environment is to decouple from it along as many dimensions as possible. When that task environment is as complex, uncertain, and unfinished as in ecosystem management, the persisting pressure to decouple means that the small scale will always have a positive role to play, though that role is never assured.

Table 6.2 Negative and Positive Features of the Small Scale

Feature	Negative	Positive
Duplication	Wasteful overlap, as many small-scale sites undertake the same activities	Redundancy; what works in some small-scale sites becomes a reservoir of innovation for other sites and localities
Fragmentation	Diffusion and dissipation of authority through many small-scale sites	Functional and spatial specialization; what works for some small-scale sites can only work there
Conflict	Many disputes, some literally over turf, because too many small-scale sites compete for too few resources	Guaranteed taking into account of different interests; voting with your feet
Polarization	Not speaking as one overall, large-scale unit to the wider public	Transparency of issues specific to the small scale, keeping tradeoffs in the wider public arena
Unintegrated priorities and goals	No large-scale, overall priorities and tradeoffs	Better protection of vulnerable goals and interests at the small scale
Accountability	No accountability for overcharging, large-scale performance	Accountability more directly related to specific tasks at the small scale
Disjoint information	No overall coordination of information gathering and assessment at the large scale	Small scale means more error correction, less distortion in scaling up information or in scaling down from the large scale
No comprehensive approach	Loss of problem-solving focus and capacity at the large scale	Decreased turbulence in task environment because small scale is comparatively more tractable
Policy change	No one in charge of the larger scale system, thereby unable to achieve necessary overall change	Incrementalist, goal-seeking change that is site-specific and tailored to local circumstance
Complexity	Disabling complexity (chaos and disintegration)	Enabling complexity (pluralism and confederalism)

The reason why the positive is never automatic lies in the coupling–decoupling–recoupling dynamic. Small-scale decoupling has to be matched by recoupling, and it is here where the large scale enters into the analysis. Much of the discussion of the CDR dynamic focused on the recoupling of activities at the operational, field level, where activities reflect issues that, in fact, are coupled at the policy level, though decoupled at the programmatic, agency, or professional levels. Indeed, the dynamic explains why "scaling down" from policy to site is often so difficult in zones of conflict. Yet, as we saw above, feedback from operational recouplings can and does change policy and program when it comes to meeting the twofold management goal. It is now time to make the complicated "scaling up" to the large scale through feedback more visible, for feedback from recouplings is clearly large scale in many of its potential implications. But how does this feedback upwards actually work? After all, achievement of the twofold management goal in zones of conflict—of which this planet has many—requires case-by-case management whose results are difficult, if not impossible, to generalize beyond their specific site. How can we have scaling up without generalization?

The best way to understand the large-scale implications of operational recouplings is to untangle the positive features of the small scale. That is, figure out what redundancy, specialization, transparency, greater error detection, accountability, less turbulence, more incremental learning, and the like offer the ecosystem decision maker at the small scale and over time. In answer, these small-scale features help in the following ways:

- to build the credibility of ecosystem decision makers, for the next stage of trying something on a wider (e.g., more intensive) scale;
- to allow ecosystem decision makers to learn by doing (or, in the case of adaptive management, actually to employ the experimental design) so as to improve management decisions with respect to the twofold management goal;
- to incrementally build ecosystem resilience as defined by meeting the goal over time;
- to enable decision makers to better detect and interpret surprises;
- to contain unexpected, highly consequential, or cascading errors that arise from the intervention (or experiment) over time;
- to clarify for decision makers when the small scale is the large scale right from the start (where and when services and functions actually are recoupled in mutually supportive ways is precisely the scale—small or large—at which the decision makers want to operate).

The credibility, learning, cumulative improvements, better error detection and interpretation, containment of the unexpected, and scale-sensitivity associated with recouplings at the small scale over time all appear to be planks in a platform on which decision makers could extend their management goals in new ways. More formally, success in recoupling at the small scale would seem to lay the foundation for extending the management goal, whether through time, space, or both. Indeed, many researchers and decision makers consider this to be what is meant by scaling up micro results to the meso and macro levels. The problem, as we have just noted, is that as much as decision makers would like to learn to work this way, they are operating in zones of conflict, where what works locally and syncretically must not be expected to work elsewhere and synoptically. How, then, do these small-scale recouplings and their positively associated behavior provide feedback to the larger scale? Again, how do small-scale sites of activities "add up" to the large scale, without the small scale being representative or generalizable at the same time? Three factors are operative here: triangulation, the coevolution of both the small and large scales, and the ability to have larger-scale impacts without having to work at the large scale.

First, we are in zones of conflict, where case-by-case management is appropriate, which means the triangulation of small-scale findings on possible implications for wider programs or policies. The decision maker seeks to determine if and the degree to which, no matter how different the sites and the recouplings, they converge on common understandings important for the operation of programs and policy. In this way, "scaling up" is a form of meta-analysis, the findings of which may have programmatic or policy implications. In fact, the meta-analysis establishes a *de facto* network or networks among the sites used in the triangulation.

Second, the CDR dynamic has a feedback cycle that takes place across scales. For example, the small scale and large scale can co-evolve together. Decision makers anticipate potential recouplings to be "out there" waiting to be identified. Yet reality in the field confronts them with the difficulties inherent in these initial expectations. They end up understanding better that the recouplings and settlement templates that work best are those dependent on the tug back and forth of all manner of local, practical, and otherwise contingent considerations. Such considerations, in turn, necessitate a rethinking of programs and policies that gave rise to the initial expectations at the outset (e.g., decision makers may conclude as a result that there should be regional or state-specific policies that should be governing the management). In another example, the feedback cycle could lead to reconsidering what the boundaries of the large scale are or should be. In the Columbia River Basin, ecosystem decision makers stressed that their planning efforts were only for federal lands and thus, ecosystem-wide management would not incorporate private property, where many of the management problems actually were. In this case, the sheer size of the Columbia River Basin planning area underscored the need to take private property owners into account for planning purposes and to come up with viable working relationships with them in the future, including through habitat conservation plans (HCPs) and other county planning fora. Here the feedback enabled the setting up of small-scale activities or partnerships addressing the private lands.

Third, the decision maker can scale up from small-scale sites to larger scale interventions without the decision maker having to manage the large scale as a whole. In our case study on the Green Heart (chapter 7), the Netherland's National Ecological Network manages many discrete nature reserves and recreational areas. Currently, the goal of the network is to expand the number and size of its sites. Our proposal provides a settlement template that extends the recoupling efforts in the sites of the network to the many small-scale activities of other organizations, such as project developers and municipalities, working in the "nongreen" parts of the Green Heart. The template brings about bottom-up coordination among small-scale sites (public and private) in the absence of an overarching organization with the mandate and authority to implement large-scale interventions. In this way, small-scale sites will have larger scale impacts without being implemented at the large scale.

One inevitable effect of these three factors—when the large is the triangulation of the small, when the small and large coevolve over time, and when large-scale impacts come about without large-scale interventions—is that an ecological network is never the sum of its parts, not because it is more than the sum, but because the small and large scales as well as the real time and long term interact with and, in fact, are inseparable from each other. Thinking over the long term requires thinking at the large scale, so the literature tells us. When the three conditions mentioned above are met, however, the decision maker can think long term from the small scale by virtue of the fact that large-scale considerations are ever-present in the small scale. Similarly, the literature argues that acting in real time is most manageable at the small scale, but when the three conditions are met, the decision maker can act in real time at the large scale by virtue of the fact that small-scale considerations are ever-present at the large scale. Thus, we would amend the environmentalist's nostrum, "Think globally, act locally," to "Think long term from the small scale, act real time from the large scale." It is just this focus, we would insist, that should set the networks of settlement templates apart from other networks of sites, localities, and management initiatives.

Now that we have developed an understanding of bandwidth management as a strategy to recouple functions and services in zones of conflict through settlement templates, we are in a position to translate this conceptual framework into organizational proposals. We focus on CALFED, our most in-depth case study.

Bandwidth Management for the San Francisco Bay–Delta

How then would bandwidth management actually work in the CALFED context? Below are two proposals that focus on CALFED's Ecosystem Restoration Program (ERP) and the Department of Water Resources' Operations Control Office (OCO) responsible for the State Water Project. The first proposal, to create a separate ERP agency, is to develop settlement templates and their networks for improving CALFED's capacity to identify tradeoffs, set priorities, and recouple bandwidths. The other proposal is to establish an "ecosystem operations" branch within the state's major HRO for water supplies, the OCO. The latter proposal aims to improve operational recoupling when managing within the bandwidths. While specific and detailed (though probably not detailed enough for some), the proposals for setting bandwidths and managing within them show how, in more concrete terms, the process of bandwidth management can be enhanced organizationally to improve the recoupling of functions and services. If implemented jointly, the proposals bring the long term and large scale together with the small scale and real time under one framework.

Establishing ERP as an Agency

A number of reasons have been given for creating a new structure around CALFED's Ecosystem Restoration Program. For one, environmentalists and others have long desired a separate unit that truly practices science-based management, subjects itself to peer review of interventions, operates under a dedicated budget for research, and has an inviolate budget for monitoring. We support the creation of a such an agency around ERP, but for reasons that follow from our framework and case evidence. ERP should be decoupled from the CALFED Program structure to enable the development of settlement templates and their network; that is, it should be created as an agency in its own right, albeit under the supervision of an interagency oversight committee.

In order to achieve the recoupling of functions and services over the long term, CALFED needs to ensure that settlement templates are being developed—if only to articulate the crucial tradeoffs, so that the political debate around CALFED revolves around the issues that matter. Given that many of CALFED's hard choices still lie ahead, the program needs a better sense of tradeoffs and priorities. Settlement templates would help it to consider these matters intelligently and practically. Furthermore, ERP would advance a long-term, whole-system perspective by connecting these templates in a network of functions and services at sites across the landscape, many of which would be small scale. Because the redefinition of functions and services—as part of the recoupling mandate—will undoubtedly include critical assessment and redrawing of the system boundaries, we advocate that ERP be set up as a statewide agency. When the system to be restored includes the water system, then ERP would be involved in projects in urban southern California as well as in the rural Central Valley and the natural areas of the Bay–Delta.

We argue that the current organizational structure is not in a position to develop the settlement templates and network. A separate agency is needed whose primary mandate it is to develop them and work on the interface of services and functions—an interface at which professional blind spots are crucial and undermanaged points. Separating ERP from the CALFED Program structure means decoupling long-term ecosystem restoration and management from the environmental activities of other agencies. A number of state and federal agencies (namely those in CALFED) already do "environmental restoration" or ancillary activities. For example, the California Department of Fish and Game (DFG) is mandated to protect and conserve the state's natural resources and the state's Department of Water Resources (DWR) has the mandate to ensure that their levee improvement projects also result "net habitat improvement."

At first it may seem that such mandates put these agencies in an excellent position to develop settlement templates. In practice, we find the reverse. The need to decouple ERP from these agencies arises from the current asymmetry, in which agencies support the importance of a long-term, ecosystem perspective for restoration (actually rehabilitation), but their management is dominated by all manner of short-term, site-specific, and urgent considerations. For example, the need to comply with stringent and complex ESA and DFG requirements during levee improvement projects leaves little room to develop and implement a longer term ecosystem perspective beyond the levee site.

Yet if an interagency program does not have the mandate, staff, or propensity to really develop a whole-system (large-scale), long-term view, why not change those mandates? For example, since the DFG is mandated to protect and conserve California's natural resources why not, as one DWR official recommended,

> expand that mandate (and provide appropriate staff and resources) to include restoration and enhancement of appropriate ecosystem components? DFG, as the responsible agency charged with [the] lead, could work with other state and federal agencies to carry out specified actions. Selection of specified actions and development of 5- to 10-year, long-term plans could be guided by an oversight committee comprised of stakeholders to ensure decisions are not left solely in the hands of state or federal agents.

But even here environmental restoration would be coupled with secondary (inter)agency considerations. For reasons already outlined, such coupling would dilute the focus and disperse the attention on the environment that would come from a separate agency. Fish and Game deals with fish and game, and it should do what it does best. Systemwide environmental rehabilitation deserves the same attention.

Our case material points to an even more compelling reason to create ERP as a separate agency. A recurring theme throughout this book has been the debilitating tension between habitat restoration and species preservation and between short-term and long-term risks. The latter, most clearly represented by the ESA, hampers attempts to implement adaptive management and wider ecosystem rehabilitation. This tension is not going to disappear. Combine this finding with the fact that developing settlement templates and their networks, by necessity, means facing difficult choices and priorities regarding the tradeoffs we identified in the previous section: between different species, between aquatic and terrestrial ecosystems, between species and other ecological parameters, and between short-term and long-term risks.

The resource agencies would find it very hard, if not impossible to make these choices and set the priorities that are needed to develop the settlement templates. Their mandate and the numerous regulations they are enforcing simply do not allow for it. When facing the tensions mentioned above, their current mandate would force them to come out on the end of species and short-term each time—as confirmed by the case studies. As we said earlier, if species disappear, no amount of restored habitat is likely to turn that event into a success for the regulatory agencies. Their accountability is framed foremost in terms of the well-being of listed and near-listed species. Simply expanding their mandate will only serve to internalize these wrenching tensions in the organization, thereby severely increasing internal goal conflict and the potential for cognitive dissonance.

Developing settlement templates is a vulnerable task, both in terms of ability to face the difficult tradeoffs and priorities involved and in terms of accountability. Vulnerability is a primary reason for decoupling ERP institutionally, so that CALFED is ensured that somewhere in the overall bandwidth management process settlement templates are being developed. With ERP decoupled, the resources agencies can still argue that they are meeting their mandate and trying to save all listed species, while CALFED management as a whole has a much better sense of tradeoffs and priorities. To put it differently, an independent ERP would capitalize on opportunities illuminated by the blind spots, most notably looking at the interface between services and functions more than any of the current organizations. This would put CALFED in a much better position to achieve the recoupling of functions and services and develop the system that can be saved.

Setting up an independent, autonomous state agency involves issues of authority, responsibility, legal mandates, funding, and staffing. These are not easily resolved and we cannot fully address them here. In addition to possible duplication of mandates, commitments would be needed beforehand that interagency differences over ecosystem rehabilitation and recovery that cannot be negotiated between ERP and the other agencies would have to be settled formally, if not by CALFED then by a statutory regulatory unit having this or related authority (the State Water Resources Control Board immediately comes to mind when dealing with aquatic ecosystems in California).

Despite these real drawbacks, we believe the potential benefits of an independent statewide agency are clear and present, as argued above. The long-run survival of a separate ERP agency would, however, depend on how well ecologists (writ large) make themselves indispensable to the other CALFED Program components and their line operations. Instead of ERP going to the other components and buying their time and staff to undertake joint environmental restoration efforts, one ERP goal should be for these components to come to ERP to initiate joint interventions. An unfortunate feature of extending ERP's active adaptive management framework to the other CALFED Program elements is that ERP has much more to contribute to these components than a six-step adaptive management process that ends up being six ways to fail. ERP's policy objective should be to give an ecological dimension to, for example, levee protection and water quality programs, not only to enhance ecosystem values but also to meet reliability concerns for these components. To put it concretely, the objective would be to improve the benefit–cost ratios of these programs by ensuring they have a value-added environmental component rather than by simply trying to improve the benefit–cost ratio of ERP interventions alone.

An indispensable ERP would be one operationally coupled with the other programs, agencies, and their line operators in major ways. ERP, for example, could contract the design and construction of levees and would itself be a contractor for DWR's habitat restoration efforts on levees. ERP would thus be taking the reliability concerns of DWR and the levee protection program seriously enough to become experts themselves in levee design, construction, and maintenance. It might, for example, employ its own engineers and consultants or canvass levee designs adopted by different countries, such as the Netherlands, to maintain comparable levels of reliability but within an ecosystem context (e.g., Van Eeten 1997).

Establishing an Ecosystem Operations Branch

Formalizing a separate organizational structure around ERP is insufficient, however, as its focus would be on setting the bandwidths. Our second proposal is aimed at strengthening and improving operational recoupling; that is, managing within the bandwidths by bringing environmental indicators and ecologists as line operators into the control room. Such input would enable a dynamic optimization process versus the more static one that is currently the case. Instead of setting overly narrow and detailed bandwidths, the ecologist as line operator will operationalize more general bandwidths in response to real-time data—including events in functions and services. Our analysis of professional blind spots showed that unexpected opportunities might emerge from getting ecologists and line operators to work together. Such an "ecological operations unit" would have to be set up in or close to the control room (or rather control rooms) charged with managing the high reliability mandates.

One state agency bears the larger part of those mandates: DWR's Operations Control Office. No other unit is better positioned than the OCO to ensure such mandates are met consistently and in a highly reliable fashion. Its real-time operation of the massive State Water Project (SWP) across many different localities and sites depends directly on the health of the Bay–Delta and above-Delta ecosystems, including but not limited to the Feather and Sacramento rivers and their tributaries. No other organization is better positioned to link ecosystem management and water reliability, as they both relate to the SWP. OCO manages and directs the project's overall water and power operations in close cooperation with the Bureau of Reclamation's Central Valley Project, with which it shares a joint operations center. Our proposal is that OCO (and through it, DWR) increase its ability to undertake ecosystem management in connection with the SWP. As with the preceding proposal, this one involves issues of authority, responsibility, mandates, funding, and staffing within DWR that we can only touch upon here.

Specifically, DWR should consider establishing a new branch, provisionally called "Ecosystem Operations," in the OCO to work alongside its Project

Operations Planning Branch, Project Operation Center, and the Project Operations Support Branch. The OCO already has the goal of managing the SWP's operations with environmental sensitivity. Yet more needs to be done to connect the line operators responsible for water reliability to ecosystem management requirements. The proposed Ecosystems Operations Branch ("ECO-OPS" for short) would provide that operational recoupling by translating a generalized commitment to environmental sensitivity into concrete ecosystem management interventions within a high reliability context.

We found evidence in all three cases that innovations capturing parts of ECO-OPS are already emerging. Examples are ecologists showing up in the control room as translators and the gaming exercises or tweak weeks (chapter 5 and above). ECO-OPS would build on and extend these innovations for CALFED. ECO-OPS would make the "secretive" process of regulatory standard-setting more transparent and explicit, just as it would open up water supply and quality operations to the ecologists. Though transparency might be resisted at first, the trust established among those involved in the gaming exercises shows that it can be done.

While our description of ECO-OPS so far has been tied closely to OCO, we would like to see it extended to other control rooms with high reliability mandates, most notably for flood control and levee integrity. There, too, ECO-OPS would aim to enable dynamic optimization between, for example, levee improvements and habitat protection and restoration. As the flood and levee management control rooms are decentralized throughout the Bay–Delta, ECO-OPS would work with or be contracted by the local districts to develop ways to better recouple levee improvement and habitat restoration within the current bandwidths of safety standards and natural resource regulations.

There are a number of scenarios as to how ECO-OPS would be set up specifically. We suggest only one. The proposed branch would consist of a small team of ecologists, line operators, and engineers recruited from within DWR and from the agencies on the branch's interagency oversight committee. Members of ECO-OPS could be part of a cross-training program, as suggested by one CALFED commentator on our proposal. However, it is important to keep in mind that the specific professional culture of line operators is the product of many years' hands-on experience. To evaluate the operational recouplings being developed, and to guide the identification of tradeoffs and setting of priorities, an interagency committee is recommended to oversee ECO-OPS' activities. The state-federal operations management team, set up to provide interagency oversight of the current OCO, appears to already provide this structure. ECO-OPS could be brought under the team's activities. This interagency oversight committee would also be the first forum addressed in cases where conflicting bandwidths cannot be recoupled operationally; that is, when issues are pushed up from the control rooms into the processes that set the bandwidths. Similarly, it would transfer the bandwidths recoupled in light of tradeoffs and priorities back to operations planning.

Whereas the challenge faced by a separate ERP agency would be to ensure its policies and goals are always operationalized in the field, the challenge ECO-OPS would face comes from the other side, namely, to ensure that its activities operate within a policy structure committed to coupling ecosystem health and water reliability in real, multiple ways that matter for policy. The proposed interagency committee should therefore consist of representatives from the Bureau of Reclamation (precisely because operational activities of the State Water Project and the Central Valley Project are so tightly coupled), the Department of Fish and Games, the Fish and Wildlife Service, the Environmental Protection Agency, and DWR's Interagency Ecological Program, among others. The oversight committee could be co-chaired by ERP and DWR in order to maximize operational recoupling between ERP and the newly established branch (i.e., between managing within and setting the bandwidths). In fact, the ECO-OPS branch could contract ERP to design some of its ecosystem management improvements just as ERP could contract the branch to undertake real-time ecosystem interventions along the lines and rationale already described. These operational links would make bandwidth management coherent, coupling ERP activities in setting the bandwidth and branch activities in working within the bandwidths in concrete ways.

Conclusion

There are many ways to extend this analysis. For example, it may be useful to rethink CALFED's overall structure. Currently divided into four key programs ("ecosystem restoration," "water supply reliability," "water quality," and "levee system integrity"), our analysis suggests the primary organizational division should be into two programs: one for setting the bandwidths, the other for managing within the bandwidths, each of which would consider the bandwidths of the four programs jointly. Similarly, one could reconsider the recent drive to establish habitat conservation plans through the Fish and Wildlife Service and the National Oceanic and Atmospheric Administration. Currently they are considered, in our terminology, to be examples of adaptive management within a context that takes bandwidth management seriously. According to the recent proposals for their handbook, habitat conservation planning uses "adaptive management ... to examine alternative strategies for meeting measurable biological goals and objectives through research and/or monitoring, and then, if necessary, to adjust future conservation management actions according to what is learned." At the same time, however, habitat conservation planning provides so-called "no surprises" assurances so that the plans can "determine the range of acceptable and anticipated management adjustments necessary to respond to new information after the permit is issued (United States Department of the Interior, Fish and Wildlife Service, and Department of Commerce, National Oceanic and Atmospheric Administration, 1999, p. 11486)."

Unfortunately, what is missing are the settlement templates that would couple the individual habitat conservation plans into a network of redefined functions and services. Or, put more positively, the challenge is to take the existing plans and determine what networks of settlement templates they imply or delineate. The notion of networks of settlement templates brings us to our final case study.

7 The Paradox Resolved

A Different Case Study and the Argument Summarized

From the outset, the book's chief policy question has resolved around the problem of how to manage. As this was addressed in the preceding chapters, a new question arises: what are the implications for policies governing the meeting of the twofold management goal of restoring ecosystems while maintaining service reliability? This chapter provides our answer to that question by means of a new case study. It sums up the book's argument and recasts ecosystem management and policy for ecologists, engineers, and other stakeholders.

The Netherlands Planning Controversy

The best way to draw out the policy-relevant ramifications of our framework and the preceding management insights is to apply them to a different ecosystem. The case-study approach has served us well in contextualizing management recommendations without, we believe, compromising their more general application to ecosystem management. Our analysis of the major land-use planning controversy in the Netherlands underscores the wider applicability of this book's arguments for both management and policy. What follows is put more briefly because it builds on the analysis of and recommendations for the Columbia River Basin, San Francisco Bay–Delta, and the Everglades.

Why the Netherlands? There are human-dominated ecosystems substantially different from those found in the United States, many of which are more densely populated. They have nothing remotely like "wilderness," but instead long histories of constructing and managing "nature." The

Netherlands is one such landscape. Not only is the landscape different, it is also important to note that the context for ecosystem management is set by different political, social, and cultural values. Sustainability is a much more dominant value in the European context than currently in the United States. Case-by-case management, while also appropriate for zones of conflict outside the United States, now has to deal with the fact that there is a tension between its call for case-specific indicators and the use of more general sustainability indicators in Europe. In the Netherlands, for example, sustainable housing projects are designed and assessed not only in terms of specific indicators but also in light of the "factor 20 increase in environmental efficiency" needed to achieve sustainability.[1]

In our Dutch case study, we move away from zones of conflict where relevant service provision is done by high reliability organizations (HROs) and where bandwidth management is our primary proposal, to deal with the paradox at the core of the book. In other zones of conflict service provision is not a real-time, high reliability activity and the notion of bandwidth is thus much less important (chapter 6). In these latter zones, the settlement template does not revolve around recoupling bandwidths, as much as around other interfaces between decoupled services and functions, such as the spatial planning system in Holland.

The case study revolves around those parts of the Netherlands commonly called the "Randstad" and "Green Heart" (further details are given in Van Eeten and Roe, 2000; Roe and Van Eeten, 2001). This region, in the western part of the country (figure 7.1), has similarities with the other three cases, but differences outweigh the similarities. The importance of the region to Dutch economic growth and spatial planning cannot be overstated. "The Western Netherlands has been a focus of Dutch national policymaking ever since [the late 1940s]. Key concepts are the Randstad and the Green Heart," in the words of Faludi and van der Valk (1994, p. 101). The Randstad is the demographic and economic center of the Netherlands. It comprises an imperfect ring of dense urbanization consisting of four of the country's main cities— Amsterdam, Rotterdam, The Hague, and Utrecht—along with the cities that lie between them, including Dordrecht, Delft, Leiden, and Haarlem. The ring of urbanization surrounds a comparatively open area called the Green Heart. "Open" and "green" are relative terms. Population densities here are also high, with urban centers puncturing this "heart," and there is little green about the Green Heart's major land use, which is large-scale, mechanized, and chemical-based agriculture. The Randstad city ring and the Green Heart make up the bulk of the land area of the western Netherlands, and population and urbanization densities are so high there that it has been called the Singapore of Europe (L. J. Brinkhorst 2000, pers. comm.).

Keeping the Green Heart open is a central tenet of the national spatial planning system. In contrast to the American context, the Dutch spatial planning system provides the legal framework, incentives and protection needed to further national and subnational development goals. At its core are the legally binding spatial plans of municipalities, provinces and central

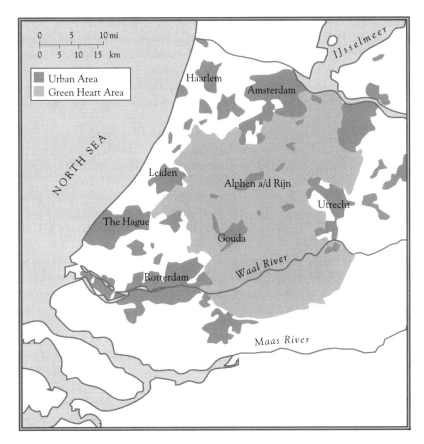

Figure 7.1 Randstad–Green Heart of the western Netherlands.

government. The current strategy to keep the Green Heart open is based on restrictive policies that severely limit the rights of local governments and market forces to develop housing and industry in the Green Heart. Formally, it is not possible for a local government to undertake development proposals that conflict with these restrictive policies. Informally, although it is very difficult, the past has seen many instances of developers nibbling away at the Green Heart. In fact, the percentage increase of urbanization in the Green Heart has been higher than that of the surrounding Randstad, such that population density of the Green Heart basically matches the national average of 440 people/km^2 (Ministry of Housing, Spatial Planning and the Environment 1996, p. 9). This is why maps such as figure 7.1 can be misleading, because their hard edges between rural and urban areas do not capture comparable population densities across the areas.

One key development area concerns the locations for new housing projects in rural and urban areas. It is estimated that nearly a million new houses have to be built in the next two decades. The twinned concepts of

Randstad/Green Heart set the framework for and determine the planning alternatives from which a choice must be made—on the Randstad city ring, outside the ring or inside the ring. Inside the city ring is rejected because of the need to preserve the Green Heart. As for building outside the ring, negative impacts have been experienced, namely, such construction increased vehicle and congestion problems in the past. So, the logic goes, the most sensible choice is to build on the ring, according to national planning policy (Ministry of Housing, Spatial Planning and the Environment 1991). More recently, the Fifth Memorandum will give stronger planning protection to the Green Heart in order to prevent the developmental nibbling away that has happened in the past (Beatley 2001).

Current policy entails major interventions and high costs. As for housing, residential construction on the Ring is much more expensive than in the Green Heart. These higher costs could translate into not enough housing being developed, thereby ensuring a persisting shortage. Another example of the costs entailed in the current protection strategy is an estimated 500 million dollar infrastructure investment for an underground tunnel that allows the high-speed train to pass underneath 7.6 km of Green Heart agriculture.

The Land-Use Problem Defined and Redefined

There are major land-use conflicts between the Randstad and Green Heart and they are by and large defined in zero-sum terms. Under current legislation and policy, the three main categories of land use—urban, agriculture, and nature—can only be developed at each other's expense. In these terms, further urbanization can only threaten the Green Heart, particularly its agriculture (the dominant land use) and its scattered nature reserves. Decision makers fear a tragedy of the commons scenario of metastasizing cities eating away at the Green Heart (e.g., Alexander 2001). Nature areas, for their part, also make claims on the countryside. The protection and expansion of the nature reserves is usually done through the Dutch government's National Ecological Network, which seeks to purchase agricultural land in the Green Heart, sometimes in direct competition with urban areas looking for expansion. From this perspective, agricultural areas in the Green Heart are seen as buffering nature reserves from the creeping urbanization of the Randstad, much as the Everglades Agricultural Area buffers the National Park from south Florida's urban sprawl. Green Heart agriculture, however, sees all such claims on its land—be they from urban or for nature areas—as a threat to the viability of agriculture. So, as currently defined, the picture in the Randstad cities can only become bleaker, as they are expected to become ever more overcrowded, nature can only become better by reducing open space, and open spaces are seen as becoming degraded when urbanized or having to produce more with less land. The need for reconceiving these zero-sum land-use conflicts is thus palpable.

Our framework provides just such a reframing. The Randstad and Green Heart of the western Netherlands constitute a zone of conflict arising out of increasing population pressures, intensified resource utilization, and the rising demand for more environmental amenities, including nature and recreation areas. There are huge population, resource, and environmental pressures in the competing land-use areas, with tensions over how best to use the land and related resources. Moreover, these tensions often revolve around ecology and reliability. The mid-1990s priority plan on rural development of the Dutch Ministry of Agriculture, Nature Management and Fisheries (1995) addresses the "tension between ecology, economy and technology," where the Ministry's "policy strives for a better balance between economics and ecology." It concludes, "This is only natural for a government that wishes to be predictable and reliable" In such zones of conflict, issues of policy coupling, agency and program decoupling, and operational recouplings of ecological functions and services come to the fore, which is certainly the case in the Netherlands. This book has no better example of coupling at the policy level than that of the interconnected Randstad and Green Heart. Planning for the Randstad continues to mean planning for the Green Heart and vice versa.

Also important, planning and implementation for urban, agricultural, and nature areas has been fragmented (read decoupled) across ministries, governments, and programs for decades, raising the ever-present need to integrate and better coordinate (read recouple) operations on the ground. Planning and implementation for a better environment is a priority precisely because the environment can only get better in an area at the expense of making the environment worse in other areas: increasing the amount of land devoted to nature areas entails increasing rates of Randstad urbanization, Green Heart agricultural intensification, or both (read recoupling ecological functions and services in one area means decoupling them even more in other areas). In this way, the present land-use problem—keeping urbanization to the Randstad, ensuring that the Green Heart remains open, and somehow expanding nature in the process—is better reframed as one of meeting the twofold management goal. The challenge is to recouple improved functions and services across a landscape that includes urban, agricultural, and nature ecosystems. Such considerations have profound policy and management implications, which we draw out through a short proposal.

Implications and Proposal for the Netherlands

The twofold management goal means recoupling functions and services not just in nature areas but in urban and agricultural areas as well. Equating "nature" generally or "ecosystems" specifically to nature reserves in the National Ecological Network is far too narrow an identification. Current nature management is operationalized in terms of how many hectares of new land can be added to the National Ecological Network's nature reserves rather than in the broader ecosystems found in the Netherlands. Ecologists, as we have seen in earlier chapters, increasingly see cities as urban ecosystems, in some cases having considerably more biodiversity than commonly supposed (Kloor 1999), while agricultural land itself is part of agro-ecosystems and agro-landscapes with functions and processes that extend beyond any given field or pasture to the countryside as a whole (Daily 1999). Accordingly, greening nature reserves by better coupling functions and services at these sites only partially meets the twofold management goal. There must also be in place a coherent policy and management structure for the greening of agriculture and cities at the same time. Instead of creating hard zero-sum edges between different land-use categories, the challenge is to treat them as a network of interlinked ecosystems over a larger landscape in terms of improving their ecological services and functions.

Many initiatives and public and private organizations are seeking to recouple functions and services (e.g., ecosystem restoration sites and incentive-based programs rewarding farmer protection of wildlife on farmlands). Unfortunately, they are typically ad hoc and within the hard edges of urban, agricultural and nature areas. While these projects are useful build on, they need to be integrated and reinforced through a different policy structure. From the standpoint of our framework, decision makers who are worried about the urbanization of the green areas should also be concerned about the greening of urban areas. Drawing the boundary around the wider ecosystems of cities, farmland, and reserves means that the greening of any one must be seen in the context of greening the others, where ecological functions and services revolving around clean air and water, for example, are as manifestly important for cities and agriculture as they are for nature areas and where the three are just as manifestly interrelated. The complexity of the overall system considerations increases in redrawing the boundary this way and thus too do the policy and management challenges, but a whole-system view also increases the management and policy options to do something different and potentially more useful than the current stalemated strategies.

How would this work in the Netherlands case? A good place to start is with the National Ecological Network. The network remains chronically underfunded, with limited revenues generated through sources such as recreation fees from existing nature reserves. Funding for anything as ambitious as the acquisition of new land for expanding the reserve system necessarily requires outside government funding or cross-subsidies from the wealthier urban and agriculture sectors, both of which are understandably reluctant sources of funding. There is, however, little reason this should continue to be the case, once new system boundaries are drawn to better reflect that the ecosystems in question include not just nature reserves but also agriculture and cities.

Imagine the National Ecological Network transformed from a conservation agency into a service agency, which would, in addition to managing nature reserves, work with other sectors and government units to rehabilitate their terrestrial and aquatic resources in such as way that ecological functions are improved from within a whole-system perspective. As a service agency, the network would not only develop and manage nature reserves, but also provide consultation services, project appraisal, post-implementation evaluation, and in some cases actual contract supervision for development proposals around new services that advance also the greening of urban and agricultural areas. Along the lines of our Ecosystem Restoration Program proposal for CALFED (chapter 6), the National Ecological Network would not directly execute development projects other than those for the nature areas, or unless otherwise contracted to. Rather than being a super agency with a massive budget, a type of proposal we criticized earlier (chapter 2), the network would have a budget sufficient to be self-supporting in its primary activities of filtering different development proposals undertaken within the spatial planning system. In short, the network would be an important adjunct to that system when it comes to the greening of the urban, agricultural and nature areas, rather than being a substitute for spatial planning and it prerogatives.

Such services would go well beyond the kinds of mitigation schemes presently in place, where urban expansion is permitted on the condition that new nature areas are created in response—schemes that reinforce the hardened zero-sum relations between nature, agriculture, and urban areas. Bandwidth management of an interlinked network of sites would now be the objective, where setting and learning about the bandwidths and adjustments within them at one site are connected to doing the same at the other sites.

Assume for the purposes of this proposal that the network is operating in a part of the western Netherlands where there are three nature reserves in relative proximity to each other. Around and between them are agricultural land and a nearby urban growth center. The current approach to nature management is to buy up surrounding agricultural land in the hope of either adding to the nature reserves or creating corridors connecting them. Species connectivity could thus be established where now there is only fragmentation and barriers to such connections. Our framework suggests a very different approach. Greening of the agricultural lands or urban areas adjacent to the three nature reserves has merit in and of itself, apart from whatever effect it would have on the nature reserves. The greening of agricultural land could take place through a number of options. A housing settlement could be located between two nature reserves, the profits from which could be used to purchase even more agricultural land around the housing estate which could, in turn, be improved so that ecological functions are now better off than they were under agriculture. A different combination of nature reserves could be linked by an office complex, the profits from which would be used to purchase adjacent agricultural land for resale only to those who would use it for more environmentally friendly or sustainable agriculture purposes (e.g., the land would be resold under the condition that it be used for organic agriculture or incorporate soil-enhancing crop rotations). A third set of nature reserves could be connected by greening the areas around the roads, power lines, and phone lines between the reserves, using money from the sale of land in the prior two cases. In this way, the network would end up not only with land uses greener than those found now on the agricultural land in question, but also with corridors between them that probably could not have been funded otherwise (i.e., the network cannot compete against the price urban users would pay for the land in question).

To undertake such greening, the National Ecological Network would mobilize experts on, for example, sustainable housing and office space, together with others into teams to develop a full service design for improving ecosystem functions within a whole-system perspective for the areas in question. The expertise could include not only better ways of managing nature but also of *reducing* the amount of agricultural land in the Green Heart, while at the same time *increasing* agricultural production, urban intensification (such as housing), and the amount of land devoted to other rural purposes there in greener ways. Ad hoc greening efforts are already under way in urban and agricultural areas (e.g., Beatley, 2000, 2001), though outside a framework for recoupling improved function and services across the landscape. Taking a systems perspective would be the specific mandate and value added of the transformed network in its service provision activities, revenues from which would be used to run and administer the network of nature reserves. In this way, the network's moving from underfunding to self-financing becomes a real possibility. Budgets will always be limited, and good planning requires decision makers to come up with ecologically sound financing scenarios for human services that do not depend on the unlikely condition of sufficient budgets to turn a significant portion of the Green Heart into corridors, nature reserves, and recreational areas.

Such an approach is not a hidden cross-subsidy, where greener housing and office settlements finance the development and maintenance of nature reserves. For in transforming the network from a conservation program into an ecological service agency, the very notion of "nature area" would itself be redefined. In terms of our framework, increasing the ambit of the network to include urban and agricultural areas as well as nature reserves necessarily moves the network's management to the right of the gradient, thereby redefining the services and functions to be managed. For example, the network would now have nature sites located in the Green Heart, say, as organic farms, or in the Randstad, say, as a business park subscribing to industrial ecology principles, a community garden initiative, neighborhood environmental initiatives, or ecological housing projects (such as between Leiden and Oegstgeest) that adopt innovative technologies to collect, treat, and recycle water instead of sending it off through sewers to treatment plants. In the process, the transformed network becomes a system of small-scale sites across the western Netherlands along the lines discussed in chapter 6.

The cynical response is to ask, do you really expect to be able to improve ecological functions through urbanized office blocks and housing estates? Do you really believe you can make the Green Heart greener by urbanizing it? The answer is yes. It is easy to see why when we realize that the baseline to which this development is to be compared is the current agricultural land use and its associated ecological functions. We know that agriculture in the

Green Heart suffers from low biodiversity and high soil and fertility losses, to name a few of its problems. From this baseline, it is plausible to improve upon these conditions while at the same time urbanizing these areas in ways that ensure the Green Heart is made greener as a result.

The greening, of course, is not automatic. The development proposals made to or by the National Ecological Network must be evaluated and supported in terms of how they would actually improve ecological functions. Proposals put forth for sites by developers, be they private, governmental, or through joint partnership, would be assessed and approved in terms of their demonstrated contributions to rehabilitating ecosystem functions. Should government continue to limit urban development in this region to a specific target, it could set this up as competition to be administered through the network between different proposals that all try to improve ecosystem functions in meeting the target. The proposals could, for example, be evaluated by an independent science and technical review committee to ensure the validity and reliability of the ecological claims made in the proposals. The core feature of the assessment and approval process would be to make explicit just what are the ecological functions in need of improvement, given the current problems in the area in terms of soil erosion, biodiversity loss, air quality, or water purity. They would also be appraised in terms of other principles that are important national spatial planning goals, such as keeping urbanization contained within fairly compact cities. These criteria may be competing, thereby forcing tradeoffs and priorities (more below). The revised version of the restrictive policies might give rise to very different urbanization patterns from those currently promoted through spatial planning.[2]

Such proposals for Green Heart housing and office blocks do not necessarily run counter to the principle of compact cities. Yet, our approach to greening is orthogonal to the current strategy, most notably the contesting urbanization at every juncture in the Green Heart.[3] Not only is the current strategy a losing proposition (urbanization in the Green Heart continues as do the number of proposals to accelerate it further), but it also fails to sufficiently recognize the degree to which ecosystem functions and services are being harmed under current land-use practices and the high opportunity costs of not taking action now in light of the real potential for improving ecological functions. Under our approach, the land developers would pay for the improvement of ecological functions as core to their overall development proposals. This would happen precisely because the transformed National Ecological Network would now be the country's service agency for whole-system improvement of ecological functions and related services. In case it needs saying, the transferability of learning about how to improve functions across and within urban, agricultural, and nature areas is optimized in such a strategy. As a service agency, the Network could still choose to invest its funds in expanding nature reserves by meeting its goal of adding 250,000 hectares to such reserves. Or, it could assess whether these funds, in whole or in part, would be better used in terms of net benefits by greening Utrecht, for instance. After a point, we may even see a change in people's very perceptions about urbanization or agriculture in terms of how green they really can become.

Such transformations require a robust policy structure that promotes the greening of urban, agriculture, and conventional nature areas simultaneously and couples their respective greening in the form of a network. Indeed, undertaking a major reform to rehabilitate ecosystem functions throughout the Randstad and Green Heart now rather than later may well be the best way to protect the region from undoubted population growth in the future. Population will continue to grow in the western Netherlands and, in terms of our approach, the rightward movement along the gradient entails increased pressures to decouple services and functions along with increased pressures to recouple them in new ways over the region. Restoring ecosystems now so that they can provide more services later is good public policy, should there be no viable way of stopping population increase in the region from occurring. These considerations thus raise the following question: how do our proposals add up to a settlement template or templates?

The Proposal as a Settlement Template for the Randstad/Green Heart

As a reminder, any settlement template is a set of redefined ecosystem functions and services within a policy structure that recouples them and in so doing erects a barrier to decoupling. The above proposals have both the elements of redefinition and structure. The redefinition starts with redrawing the boundaries around cities, agriculture, and nature, which leads to recasting the "greening" of functions and services as a form of ecosystem rehabilitation of the Randstad/Green Heart. The revised set of functions and services that results is pulled out of zero-sum into a more complex, richer set of tradeoffs and priorities across the ecosystem, thereby seeking to resolve the paradox of rehabilitating the Randstad/Green Heart while also providing improved and new services, such as housing and office space, among others.

The policy structure resides in the fact that (1) the developers have strong incentives to optimize functions in their proposed investments in services and (2) the transformed National Ecological Network has strong incentives to optimize services when approving or supporting the investments in terms of their contributions to functions. Our proposal, like the accounting structure of the Environmental Water Account (chapter 6), provides a mechanism which ensures that every investment proposal for services is coupled to an investment in functions that can be traded off against other options (i.e., greening the Hague versus greening the Green Heart through, e.g., restoring wetlands in the latter). There are incentives on both sides to optimize the other. On the services side the incentive to optimize functions comes from the fact that the developers need government permission to undertake their proposed investments in the Green Heart area. Since development in the latter is severely restricted, only those proposals that optimize functions most effectively are likely to be supported. On the functions side, the incentive to optimize services comes from the fact that the investment proposals actually need to make economic sense. The more the proposals are economically feasible (both in number and amount), the greater the potential investment base for ecosystem improvements promoted by government. This policy structure becomes visible only when ecosystem function and services are redefined out of zero-sum relations.

The implications of having such incentives are also profound. Here we mention a major one. Chapter 4 pointed out that the trinity of better prices, institutional reform, and incentives is an insufficient solution to improving ecosystem management without a wider framework that confirms which ecosystems would benefit from which interventions in meeting the twofold management goal. Now that we have the framework and a proposed settlement template for the Randstad/Green Heart, the form these incentives would take in a concrete situation is much clearer for the decision makers concerned.

Resolving the Paradox: The Argument Summarized

In truth, we could continue to explore the policy implications of our framework, and our hope is to do so in future publications. We could also explore further the similarities and differences among our four case studies.[4] That said, with the fourth case in hand, we are now in a position to sum up the basic arguments of our book.

None of the terms in ecosystem management are precise or clearly defined, but all the underlying issues are hard and clearly persisting. The buzz terms will change (today ecosystem management, tomorrow something else), but the challenge of reconciling people's demand for reliable services from the environment (including clear air, water, and power) with their demand for a clean and safe environment that requires rehabilitation is a challenge that is not going away any time soon. *Ecology, Engineering, and Management* uses ecosystem management to explore these wider, seemingly intractable tensions and to find out how to better address them in future.

In our view, the goal of ecosystem management is to recouple ecosystem services and functions in such ways that (1) where decision makers are managing for reliable ecosystem services, they are also improving the associated ecological functions and (2) where they are managing for improved ecological functions, they are better ensuring the reliability of the ecosystem services associated with those functions. Improving ecological functions covers a range of activities, including but not limited to preservation, restoration, and rehabilitation. It typically means (re)introducing to ecosystem functions a degree of complexity that they may have had historically, or may need now, in order to maintain keystone species and settlement processes. Thus, the challenge of ecosystem management for both ecologists and engineers (writ large) is to improve ecosystem functions while at the same time maintaining, if not improving the reliable provision of ecosystem services. How then are decision makers to preserve, restore, and otherwise rehabilitate ecosystems and their wider landscapes, while at the same time ensuring the reliable provision of services (including goods and commodities) from these ecosystems? This book's answer has several parts.

Matching Ecosystem and Management Regime

First, the decision maker matches the ecosystem of interest to its suitable management regime. Ecosystems vary across a landscape in five interrelated ways, each important for the purposes of their management: human population densities range from low to high; resource extraction ranges from virtually nonexistent to intensive; availability of causal models to explain and predict relationships important for management varies from few and inadequate to many and more adequate; ecosystems can also vary in the degree to which ecosystem health or organizational health considerations predominate in management; and ecosystems differ in the degree to which they represent a source of multiple functions with few services to those whose resources each provide multiple services.

Along this fivefold gradient moving from few and low to many and high, four major management regimes of mutual interest to ecologists and engineers are identified for the purposes of ecosystem management (figure 4.1): self-sustaining, adaptive, case-by-case, resource, and high reliability management regimes. At the left end of the gradient, there is self-sustaining management for those ecosystems with relatively few people, low extraction, few causal models, high priority given to ecosystem health over other considerations, and with few human services provided from a highly complex, holistic ecosystem. At the gradient's other end is high reliability management for those ecosystems characterized by relatively large population and extraction levels, many causal models available for use in management, higher priority given to organizational considerations typically mandating reliability of ecosystem services, and by the demand for and provision of many services associated with individual natural resources rather than the ecosystem as a whole. Between self-sustaining and high reliability management are the adaptive management and case-by-case management regimes, to which we return shortly.

Each management regime differs not only in terms of gradient's five dimensions, but also in terms of beliefs about the ecosystem being managed, types of models used in the management, modes of learning about the ecosystem being managed, stakeholder involvement, and measures of management success and failure (table 4.3). Especially important are the learning thresholds that separate each management regime from the others. Limits to learning, such as the precautionary principle, are the primary mechanisms that compel decision makers to move from one management regime to another. From the perspective of our framework, the fundamental policy question in ecosystem management is, how to manage? The fundamental policy implication of the framework is that the exclusive use or recommendation of any one management regime, be it self-sustaining, adaptive, high reliability, or case-by-case management, across all ecosystems within a highly variegated landscape that is variably populated and extractively used is not only inappropriate, it is fatal to meeting the twofold goal of ecosystem management.

Redefining Ecosystem Functions and Services Along the Gradient

Even if ecosystem and management regime are matched, can the management goal be met in each case? Here is the second part of our answer to the paradox of how you improve ecosystem functions while maintaining ecosystem services. The answer follows from the fact that while all ecosystems, even self-sustaining ones, are in need of better recoupling of services and functions, the services and functions in need of recoupling are never the same everywhere. Functions and services are redefined all along the gradient and they too vary by management regime, and importantly so for the decision maker.

Redefinition of services and functions occurs in several ways. Some are under the decision maker's control; many others not.[5] As one moves rightward along the gradient, services are increasingly decoupled from functions, with consequences for what the decision maker takes to be services and functions. Once the rivers we used had floodplains; now they are channelized. Once the land we cropped was left fallow; now it is used year-round. Once the wilderness we visited was virtually self-managing; now a huge stewardship infrastructure is in place to manage the nature areas that are left. All of this has been done under pressure of increasing population and extraction. Unfortunately, these changes, which have enabled decision makers to separate services from functions, have also led to large environmental costs and increased pressure to realize the twofold management goal.

The decoupling of services and functions is driven by a wider decoupling that is taking place under pressures along the gradient. What started out on the gradient's left as a complex, holistic ecosystem ends up on the gradient's right, disaggregated into extractable resources, each having its own next-best substitute and each of which has many different uses. The whole ecosystem is decoupled into discrete natural resources. Ecosystem functions shift from a dynamically interactive whole with few "natural" services into specific functions coupled to increasingly specific services. Self-sustaining functions move from many to few, if any, while services move from natural to highly discrete ones.

Decoupling resources from system and services and functions from each other has profound implications for how to redefine the services and functions in order to recouple them more effectively. In the first place, ecosystem management moves from the presettlement or predisturbance template to settlement templates with goals, objectives, and scenarios to rehabilitate specific functions associated with specific services, even if it is impossible to restore the presettlement template. Preservation gives way to restoration, which gives way to rehabilitation, as the presettlement template gives way to multiple settlement templates in which the twofold management goal could be realized and networked.

Managing Ecosystem in Zones of Conflict, Case by Case

The third step is for the decision makers to realize that many of the ecosystems requiring rehabilitation and other ecosystem management improvements are in zones of conflict. Pressures on this planet are moving ecosystems to the right of the gradient. Increased population, extraction, improving causal models, more techno-managerial organizations, and the demand for ever more services from whatever resources we have are all pushing ecosystems toward a high reliability management that is, ideally, environmentally friendly rather than destructive as in the past. Indeed, such environmentally friendly, high reliability management is already evident, be it the human guard protecting an endangered elephant or the massive and far more expensive infrastructural "guards" and "fences" seeking to protect every ESA species in the United States or all of the Green Heart in the Netherlands. The stewardship infrastructure to protect, restore, rehabilitate, and otherwise maintain endangered species, habitat, and ecosystems remains, however, nascent and clearly inadequate to the task ahead of realizing the twofold management goal for *all* ecosystems.

In facing such challenges, the understandable temptation is to insist that the only truly effective way to protect and improve ecosystems is to reduce population growth and extraction. In this view, fewer people and less consumption are really the only long-term solutions to saving the planet. Unfortunately, the strategy of moving ecosystems to the left of the gradient or preserving them if they are already there treats the twofold management goal as if it were primarily a matter of reducing or redefining services entirely in terms of functions. More formally, reducing population growth and per capita consumption are strategies that seek to decouple forever what are perceived to be harmful services from the functions or necessary services they are said to endanger. In this way, such proposals mirror the very state we now find ourselves in, where services and functions are already decoupled from each other. These strategies serve only to change the mix of decouplings, because whoever gets to determine what are harmful services gets to delink them from what they consider to be necessary functions and services that remain. Many ecosystems are in zones of conflict and will remain there, where neither functions nor services override the other. The challenge persists: how do you recouple the services and functions that are there? Even if population and extraction were reduced, there would still be the need to better couple the remaining services and functions, which is the challenge everywhere along the gradient.

This is *not* to say that population growth and extraction have not had profoundly negative effects on the landscape. The environmental costs associated with decoupling services from functions and resources from the whole ecosystem have already been identified. In our view, rapid population growth and human extraction have actually left this planet with highly interconnected systems—economic, social, political, cultural, organizational, and ecological, among others. The systems are so complex and dynamically changing that the decision space in which decision makers must operate is fraught with unknown risks. Given the complexity of ecosystems and the multiple scales over and within which they operate, errors and surprises are always possible in any intervention, both over the short and long run, and complicated by the fact that the errors could be cumulative, irreversible, nonlinear, indirect, and highly consequential. Added to this fallibilism in outcome (mistakes are always possible, not so much because the risks are large as because they are unknown) is the pluralism that also characterizes the decision space of the ecosystem manager. Cultural, political, and societal values are many. In some cases the consensus over what to do holds; in many others, it does not.

In brief, the decision space in which the decision maker most likely operates today is one of conflict. Not all ecosystems are zones of conflict, but many are, and in terms of our framework, the latter require case-by-case management. This book defines case-by-case management as those interventions involving various stakeholder groups and decision makers that improve ecosystem services and functions both in the field and in ways that better match the interrelationships among functions and services from a whole-system perspective. More formally, case-by-case management is the recoupling of ecosystem functions and services by decision-making units on issues that are coupled at the policy level but decoupled for the purposes of the units' decision making into separate programs, agencies, and professions. How these recouplings actually take place necessarily varies over space and time, that is, case by case.

The principal features of case-by-case management—ideographic patterns of complexity and interaction of the ecosystems to be managed combined with the evolving nature of management over time, the multiple criteria to evaluate success or failure of that management, and the use of multiple methods and sources to triangulate on what should be done by way of management—all ensure that the performance record of case-by-case management is mixed, never totally negative but never entirely positive. That is precisely why it is difficult to generalize or replicate from case-by-case management.

Nonetheless, case-by-case management is actually going on, albeit decision makers frequently mistake it for failed versions of ecosystem management. The prism of case-by-case management allows us to see different arrays of options for recoupling. It also reveals many important innovations for managing in zones of conflict that are already part of current or emerging practices, albeit misunderstood as failed adaptive management. These innovations revolve around the following: trading off scale and experimental design; integrating planning, programming, and implementation; bringing ecosystem functions into real-time management; developing comprehensive models and gaming exercises; increasing the water budget through storage.

Managing in Zones of Conflict: Bandwidth Management

Many zones of conflict are populated by organizations that demand or aspire to the high reliability provision of services, albeit in ways that are more environmentally friendly than ever before. In these situations, managing case-by-case can be thought of in terms of managing bandwidths. In other zones of conflict, service provision is less of a real-time, high reliability activity. There, case-by-case management is less focused on bandwidths. We discuss those situations in the next section.

The common thread in the innovations just mentioned is that they overcome the professional blind spots of ecologists, line operators, species-specific regulators, engineers, and modelers that threaten the twofold management goal. Blind spots are the result of the specialized narrowness that comes with normal professionalism—the thinking, values, methods and behavior dominant in a profession. For any profession, normal problems have normal solutions, even if *that* is the problem. A crosswalk is needed between the professions, or in our case between adaptive management and high reliability management, the two regimes most prominent in case-by-case management.

Bandwidths are a primary means though which HROs maintain highly reliable operations. Some parameters are given literally in the form of bandwidths, such as regulations schedules for minimum and maximum water levels. Fish protection or water quality standards also define limits within which operators have to keep the system. Many other examples are available. What we call "bandwidth management" distinguishes two distinct but related processes: setting the bandwidths and managing within them. Managing within bandwidths happens within or around the control rooms of HROs. It is learning where the flexibility is within the current bandwidths to recouple functions and services, as we saw in the case of the CALFED gaming exercise.

Bandwidths are loosely coupled to the full range of internal goals, tradeoffs and priorities that the agencies have. They are also by and large decoupled from the bandwidths of other organizations. Both lead to conflicts between and among bandwidths. Such conflict at the inter-organizational level is addressed through the process of setting the bandwidths. In three of our case studies, the negotiation process over bandwidths is frequent and unavoidable. The bandwidths that the ESA and other regulations pose, together with the bandwidths for reliable service provision, produce overly constrained situations and infeasible solutions for line operators. The conflicts between these decoupled bandwidths is pushed outside the control rooms and fed into a process of (re)setting the bandwidths. Bandwidths are negotiated at the inter-organizational level, if only because the control rooms need unequivocal bandwidths in which to operate. The negotiated bandwidths, in turn, have an impact on the organization's own complex set of goals, tradeoffs and priorities. The nature of this impact is an empirical matter and depends on the degree to which the set is decoupled from the bandwidths. The inter-organizational process of the striking tradeoffs and setting priorities through the medium of bandwidths is the most tangible site of recoupling of functions and services we have found in the cases studies.

There are different ways to negotiate the discrepancies between the functions and services as defined by the current bandwidths. Zero-sum solutions are possible, along with win-win solutions and solutions that recast the problem, that is, redefine the functions and services in question. The redefinition of services and functions is the most fundamental way to address the discrepancies between bandwidths and, in light of the paradox, a necessary one. While it is a crucial task, the professionals involved are least equipped for it, as the effects of blind spots are most negative here.

We saw how redefinition is always going on along the gradient, not as a managed process, but as the de facto outcome of changing landscapes. While much redefinition is implicit, there are more explicit instances. New and different functions and services are discovered during the management of ecosystems. A crucial push in this dynamic comes from redrawing system boundaries, for example, such as in the Everglades, where including urban and agricultural lands in the system has led to reconceiving the agricultural area from being a threat to being a buffer against greater urbanization. Redrawing boundaries capitalizes on the positive features of the blind spots. It leads decision makers to generate new or hitherto unrecognized options for improving ecosystem functions while at the same time ensuring service reliability. It leads them to settlement templates.

Managing in Zones of Conflict: Settlement Templates

Settlement templates are the guides to recoupling services and functions that decision makers have when the presettlement template is unsuitable for purposes of ecosystem management. Instead of trying to restore a system that cannot be preserved, the settlement template defines and develops the system that can be rehabilitated and maintained.

Where there are HROs, the settlement template revolves around bandwidths. The template is a set of redefined functions and services within a policy structure or mechanism that brings more coherence to currently decoupled bandwidths. As in the Green Heart case study, there are also zones of conflict where service provision is not a real-time, high reliability activity, such as housing. In these zones of conflict, the settlement template does not revolve around recoupling bandwidths, as much as around other interfaces between decoupled services and functions, such as the spatial planning system in the Netherlands.

Settlement templates come in all shapes and sizes, and time and scale dimensions. One settlement template could bring ecologists into the control rooms of the major power and waterworks in order to improve real-time management. Another settlement template could ensure that metropolitan water districts have significant ecosystem rehabilitation and management budgets and units. A third settlement template might develop a network of ecosystem sites whose implementation of the twofold management goal synergistically connects land-use categories normally considered in competition or antithetical to each other. A fourth settlement template could combine elements of the three. A fifth template could be markets that were able to internalize externalities as a way of correcting market failures. In all these examples, there is both redefinition of functions and services for the purposes of their recoupling and a policy structure that provides a barrier against the pressure to decouple.

The more fundamental point here is that settlement templates are always multiple. They are multiple because cases are plural, scales are never singular, time horizons necessarily vary, and the services and functions are perforce many in zones of conflict. Indeed, the problem has been thinking and acting in terms of a one-size-fits-all settlement template. The core difficulty with the Endangered Species Act and similar species-protection legislation is that such laws aspire to be a form of high reliability management, whose settlement template is one function (biodiversity) and one service (species preservation) regardless of time, scale, and cost. Yet the limitations are obvious when the species under threat is mobile and when tradeoffs and priorities are called for precisely because cost is a factor. Not all the threatened species can be saved at once; recovery is rarely clear or certain; and preservation can never be the one and only answer. To act as if these limitations do not exist amounts to a colossal exercise in castle building. Turn the Everglades into a costly high-security prison for zoo animals, turn the Bay–Delta gates and canals into a members-only gated community for delta smelt and elect others, turn the Columbia River dams into massive guards for salmon on the principle of one guard for one elephant, and turn the Green Heart into a high-class museum that no one can pay to get into with a rural lifestyle on display that no one can really afford, and here you have the single-function, single-service settlement template writ with a vengeance. What you wanted in the stewardship infrastructure was a highly reliable fire-fighting unit that saves life and land; what you end up with is a theme park where you hope the rides do not fail.

Are we saying that the salmon, delta smelt, wading birds, and other threatened species should be left to go extinct? Of course not. We are arguing for a set of settlement templates able to redefine and better account for species rehabilitation or recovery. If, as we have argued, ecological services and functions are redefined all along the gradient, then the context for assessing species recovery is necessarily redefined as well. In one part of its habitat salmon is treated as a sport fish, under the settlement template operating there; another part of the ecosystem has a settlement template that treats the salmon as primarily a spawning species to be protected; in yet another part of the landscape, dying and dead salmon are considered by that area's settlement template as key to organic-matter accumulation in the river. In this stylized example, recovery of the salmon is equivalent to ensuring connectivity between the three settlement templates and ensuring that salmon have multiple functions and services within the context of the network of three sites as a whole. Salmon are what connects the settlement templates into a network of changing services and functions associated with that species, such that salmon recovery would make no sense if assessed solely in terms of a one-function, one-service template.

The implications of multiple templates require fundamental rethinking of species recovery. First, since recovery of a threatened species is recoupling a changing set of services and functions associated with that species across a settlement template network, the species never "fully recovers," if simply because redefinition of services is never final or definitive, especially in zones of conflict. Second, there are two ways to kill what recovery there is: one is by letting species die; the other is by killing off the services the species provides and the functions it fulfills. Obviously, extinction of a species makes connectivity between templates and recoupling between services and functions moot. Less understood, but equally debilitating for recovery, is what happens when a threatened species is reduced for management purposes to one priority function (biodiversity) and one priority service (species preservation). Simply put, the rationale for its management is considerably weakened. Endangered species legislation may mandate the high reliability management of certain species, but, as we have seen, such management is more forthcoming and appropriate when the resources being managed provide multiple services. Without settlement templates that define what services and functions of the species are important for recoupling and without the network of templates to show how these different attributes of the same species are connected, recovery degenerates into counting samples from unknown universes and responding as if that exercise were adequate enough.

Third, we have to radically reconsider the standard argument for the one-function/one-service/one-template approach to species preservation and recovery. In our terminology, this argument asserts that saving species from going extinct keeps open the possibility of having more or new settlement templates, services, and functions than is currently the case. Somewhere in all that threatened biodiversity, we are commonly told, may be the cure for cancer. But that is precisely the point, isn't it? Where are the real-time management plans for saving biodiversity as a way to cure cancer? Where are the settlement templates that seek to improve ecological functions, including biodiversity, for the purposes of improving reliable services, including cancer cures or treatments? Where are the bandwidth management strategies that go beyond, say, working with *Taxus* and taxol, some Amazonian herb or the like, to set broader ecosystem bandwidths and manage within them as a way of reducing cancer? And where are the trade-offs and priorities between such settlement templates and competing ones? It is simply not enough to say that saving species keeps open possibilities for saving the future. The future, like the present, is never one way only and what will secure these alternative now-and-thens—the settlement templates—is hard work and luck, not possibilities.

Here is how the paradox is resolved. Settlement templates—a set of redefined and recoupled ecosystem functions and services within a policy structure that creates counter-pressures against decoupling—convert the paradox of the seemingly mutually exclusive improvements in functions and service reliability into a relationship between improvements that is non-zero sum or even mutually advantageous. The decision maker ends up with a revised set of functions and services that has fewer zero-sum relations (or biophysical decouplings), thus enabling a dynamic optimization of functions and services.

Networking the Small and Large Scale

Networks of small-scale sites can be at the heart of any given settlement template, such as we discussed for the Dutch National Ecological Network and the proposed Ecosystem Restoration Program of California. But there are also networks of settlement templates. For both, an important issue is that of scaling up and down.

While it is not possible to generalize from one case to another in zones of conflict, it is possible to take the next step and scale up or down what has been learned from case-by-case management in these zones. There can be no guarantees here, since the task is to draw ecosystem boundaries in ways that encourage decision makers to think about long-term improvements in services and functions from a small-scale perspective and to undertake these improvements in real-time from a larger scale perspective. Functions are redefined because processes invisible at one scale are revealed at a larger one (e.g., hydrological and flow transfers between sites). Services are redefined for the same reason (e.g., flows and transfers of people and commodities between sites also become clearer).

A positive theory of the decoupled small scale justifies scaling up small-scale ecosystem management interventions in ways that do not depend on the small-scale site being statistically representative or the intervention being replicable. Three factors are operative: triangulation of larger-scale findings from a meta-analysis of small-scale activities; the coevolution of both the small and large scales as decision makers learn more and more about ecosystems and their management; and the ability to have larger scale impacts without having to work at the large scale.

Funding and Undertaking More Science, Engineering, Adaptive Management, and Model-Based Gaming

Our talk of settlement templates, bandwidth management, and recoupling services and functions cannot substitute for the more and better science and engineering needed if these goals are to be realized. While ecologists differ over how ready they are to advise on habitat restoration, recovery of species, and ecosystem management generally, they do speak with one voice in calling for more research on ecosystems functioning and processes. Engineers and better engineering for ecosystems must be part of that call. We endorse these appeals and hope that our framework guides their choices of what to study and recommend. The need for more research and better engineering is urgent and requires, we believe, a national, regional, and local policy and program structure committed to the rehabilitation of ecosystem functions and keystone species central to the provision of ecosystem services. Each country's program must be countrywide, understanding of course that what works by way of settlement templates and meeting the twofold management goal will necessarily vary. A Marshall Plan for policy- and management-relevant research on ecosystems is needed, be those ecosystems urban, rural, wilderness, or other. The planet's first worldwide defense budget should be one directed to defending the environment for the people who use and derive services from it. Part of the needed research, moreover, must be in the form of adaptive management. Instead of advocating the end of adaptive management, we have recast it as part of case-by-case management in zones of conflict. Here, adaptive management's contribution is potentially most helpful but least understood.

We have seen the problems with adaptive management. Nevertheless, a landscape-wide policy framework that recognizes management is case by case, that accepts the need for a long-term perspective as much as a real-time one, and sees the positive, scaling up features of small-scale interventions above and beyond their representativeness or replicability, is one that can and must have a strong and positive place for adaptive management. Such management will not be the large-scale experiments currently touted, but rather smaller ones in those localities where long-term experiments are possible—politically, socially, and ecologically. Obviously, this will not be everywhere, but ecosystem management is about the twofold management goal, not about being the same everywhere. In this way, adaptive management experiments become an important part of triangulation and meta-analysis exercises that converge, from many different directions, on what should be done by way of meeting the management goal. When it comes to researching things as complex as human-dominated ecosystems, the most invidious distinction that can be made is to compel decision makers to choose between quantitative versus qualitative or reductionistic versus holistic approaches to such research. In the highly dynamic decision spaces that are human-dominated ecosystems and landscapes, the decision maker needs all of these methods, and more, when trying to get a fix on what to do.

Last but not least, model-based gaming exercises are a wonderful example of triangulation based on multiple databases, models, methods, and key informants. Modelers and decision makers keep calling for more comprehensive, linked models that mimic the wider interactions and interrelationships of the environment. But as we saw earlier, the gaming exercises are de facto those linked models being called for. Those attending these exercises function as the links that have so far escaped more formal modeling. The game is a simulation of system behavior, when the system is taken to include the natural and the organizational. Because the participants in the games are the links, scenarios can unfold, reverberate through the system and produce surprises from which learning can then take place. We endorse such gaming and related exercises precisely because they appear to be comparatively effective mechanisms for exploring the recoupling of services and functions within a whole system perspective for the purposes of settlement templates and in ways that compensate for the professional blind spots of the individual professionals participating in these exercises. Gaming can also be useful in other sectors, such as transportation, and there is much to learn from these other exercises for ecosystem management projects.

Coming Back to Where the Book Started ...

In Lou Reed's phrase, we are at "the beginning of a great adventure." Here are some concluding thoughts about how we ended up at this specific beginning.

We are visiting a large levee island in the Bay–Delta. We drive around the island atop the levee road, as Sally shows us her water habitat restoration efforts on the outside of the levee. Every so often, she stops the truck, we jump out, and she shows us her hard-won, water-side improvements. She tells us all of this has backfired on her. "You know what happens?" she asks us. "You save the habitat and then they have to regulate you more. Look there, the island across the channel got rid of the habitat and threw rock all over it. Because there's no habitat, there's no regulation." Now Sally points Michel to where a bund has been constructed, while Emery takes his first long look at the farm-side of the levee. It is irrigated land leveling off into the distance. Then it strikes us. We are looking at the wrong ecosystem. We are being shown the water habitat on the left, but who is looking at and asking about the agro-ecosystem on the right?

After another long drive, we are at the Dairy Queen somewhere between Boise and Portland to meet Tom, our next interviewee. We have just plugged in the laptops when we see him struggling to pull his trolley load of documentation through the door. Without taking off his coat, Tom sits down and speaks for half an hour before we can ask our first question. The first thing we see when entering Steve's office in West Palm Beach is the wall of documentation related to the Everglades restudy. Steve, like Tom, is really impressive in laying out the sheer complexity of what they are trying to do. "Here," says Tom, "take the CD"; "Here," says Steve, "take this CD. It's got all the documents in digital form." We think, not for the first time, where is the framework to pull all this together?

Steve takes us down to the control room of the water management district. Tommie shows us around. Over there are the workstations. We see some of the many spreadsheets. That side, a new intelligent warning system station is being debugged by engineers before being passed onto control-room line operators. The line technicians want to know for sure that what they are seeing is real, real-time information and not bugs in the software. Tommie says they are hiring an environmental sciences person to manage wetlands in real time for stormwater treatment. He will work with the technicians and engineers in the control room. We look at each other. This idea was in our earlier CALFED report.

We are in the control room of the State Water Project, Sacramento. Joan Didion in *White Album* (1979) describes the place as one where the "best efforts of several human minds and that of [the computer] are to be found." Emery spins out the elevator speech of our book's argument about being caught in the middle of trying to rehabilitate ecosystems and still keep the water flowing. Curtis's eyes light up. He is in the middle and appreciates we know that.

Rob is in the middle too, but does he know it? In another Sacramento office, he shows us a figure from one of his reports. The basic idea is to integrate the ecology and reliability of a levee by taking the habitat from the levee and move it preferably to the water side. A major advantage of this, according to Rob, is that it would eliminate the dilemma between conservation and levee maintenance. Now each has its own domain. They're separated and, consequently, so are the administrative responsibilities. It strikes us afterward that this is not integration. It is our first real example of decoupling.

"Afterward" is in the conference room of a technology cluster for start-ups, where it is only the two of us and a whiteboard marked with blue spaghetti. We are trying to put all of the major CALFED activities into a table. How do all they add up to meet the ecosystem and water reliability challenges facing California? In CALFED plans, everything is connected when it comes to the environment. But look at our list: it is all separate programs, units, and mandates. Coupling, decoupling. Still, what about those connections made in the control rooms, like Tommie's environmental hire working as a line operator? Recoupling. Next thing you know, we are arguing that coupling leads to decoupling and that is necessary to get anywhere with recoupling where it matters the most, on the ground. Later we see that there are more positive features of decoupling. The whole things starts to click together for the first time.

We are still working on CALFED and got ourselves invited to one of the gaming exercises. Squirreled away in a government building is a group of operators, regulators, and modelers exchanging esoteric codes, looking at monitors and screens, waiting for computer runs for the next month, hoping that their virtual populations—the fish, the crops, the Los Angelinos—will make it this time. Last round, they had run out of water before the game ended. As they muddle through more rounds, things get better. This push and pull of people, spreadsheets, and game rules turns out to be the whole-system model everyone has been looking for, including them. Later we understand why this is the most important thing going on in CALFED.

The most important thing going on in the Dutch Green Heart policy is the government's new Nota (white paper). We got a meeting with Laurens Jan Brinkhorst, the minister of agriculture and nature management, to talk about our ideas over breakfast. While the minister is spreading marmalade over toast, Michel talks and Emery takes Brinkhorst's fruit. "See this apple?"

Emery asks the minister. "That's a nature reserve. See this banana and this orange? Those are other reserves. The corridors in between are constructed and financed around sustainable office blocks and housing." Then we lay out the rest of our proposal. Later when we write up the idea for the newspaper *Trouw*, someone writes back saying that rabbits do not live in high rises.

In just the same way, salmon find it hard to live with high-rise fish ladders. In downtown Portland, Chip explains how salmon populations are dying off; the experts feel like they have tried almost everything, and it is not working. Now they want to take out dams and cut off electricity. It is the end game in which everything looks ridiculous. The hatchery salmon are too fat or they stick out like neon or they volunteer as duck fodder. Once he thought adaptive management was the sure way to go, now they need fresh ideas, which are in short supply. What is not in short supply are half-hearted recommendations that do not get us anywhere, but cannot leave us where we are.

And so we end in Leiden. We take a break from the final edits and are in a restaurant. What do we most remember, we ask ourselves. What do they need to know that we have not said already? Sally, Tom, Steve, Tommie, Curtis, Rob, Laurens Jan, Chip, and the rest—some more passionate than others, all were serious and knowledgeable, without the easy answers of outsiders. If many of their organizations have created untold environmental harm over the years, they are part of the solution—if only because ecologists, engineers, line operators, bureaucrats, modelers, farmers, and politicians are part of any solution. That is, if there is a solution to be had. To resolve the paradox means you have to treat it seriously as a paradox right from the beginning.

Appendix: Modeling in the CALFED Program

Modeling and models have been instrumental in CALFED, as in the other initiatives. For this reason, this appendix adds more detail to the observations discussed in chapter 3. The simple model underlying the X2 salinity standard helped secure the Delta Accord in the first half of the 1990s leading to the creation of the CALFED Program. The Vernalis Adaptive Management Plan (VAMP), commended as CALFED's best example of adaptive management, estimates the parameters of a biological model measuring the impacts of flow and export variables on fish. Its results are intended to have significant management implications. Engineering models of water flow are used in many ways by CALFED agencies and program elements, such as in assessing the merits of various alternatives. Both engineering and ecological models, in the words of one informed observer, "help us at the bookends" in the gaming exercise, which involves ecologists, operators, and other stakeholders simulating, over several days, real-time water allocation decisions among competing water uses and in the face of persisting resource scarcity. The gaming exercise as well as a recent spin-off modeling effort are credited with providing crucial insight and advice on how to better handle crises like that of the delta smelt in December 1999. A CALFED interviewee describes the gaming as giving "biologists a feel for the reliability issues."

While ecological models are central to adaptive management, actual practice has left much to be desired. In the words of one CALFED informant, "Clearly, there is a better suite of models for the engineers [and line operators] than for the biologists [in the Ecosystem Restoration Program (ERP)]." The asymmetries between engineering models for high service reliability in water supply and ecological models for adaptive management of aquatic ecosystems are multiple and work against the successful implementation of model-driven adaptive management in the ERP as in other ecosystem management initiatives.

First, even though CALFED goals and objectives are fairly clear and defined, programmatic and management priorities are not. This matters for the link between adaptive management and the ecological models on which they are based in ERP. Sequence was a concern of CALFED staff responsible for the ecosystem restoration component: whether to identify programmatic priorities first and then undertake adaptive management with respect to those priorities, or to first develop the broad ecological models and then undertake comprehensive baseline monitoring, upon which to base and drive the adaptive management thereafter. Both must be done, but clear and defined priorities from the outset are critical in the absence of the presettlement template to guide management. In practical terms, these priorities and the wider goals and objectives they reflect must serve as the template against which any model building and testing is calibrated.

Establishing priorities and developing models in light of those priorities are crucial for managing aquatic ecosystems. It is common to hear variants of, "but we don't even know the basic biology of the Bay." Even if CALFED ecologists and biologists had the models, the problem remains of empirically estimating their parameters. While one CALFED interviewee argues that baseline information is readily available for some species, others are less sanguine. They argue that few CALFED programs have the baselines or databases from which to estimate model parameters. One ecologist puts it this way: "Across the board, there really is very little understanding within CALFED of the massive amount of information necessary to actually do adaptive management but currently is not there or missing altogether." While adaptive management is learning just what those real-world parameters could be, it cannot proceed effectively without some reliable parameters to drive the models upon which management options are to be derived, tested, and redesigned in light of learning. Adaptive management works best, again, when the models are there already.

The horns of the dilemma are illustrated in the Vernalis Adaptive Management Plan. Over a decade or so, VAMP is to use some $48 million to estimate three parameters of a fairly parsimonious ecological model comparing impacts of water flows and exports on fish. There is no control group and no possibility of replicating the interventions. When asked, "What if VAMP research ends in only a scatter plot of results showing no discernible or firm relationship between flow and exports?" a CALFED consultant shot back, "If?" An issue in CALFED has also been how to peer review models that are, by their own modelers' estimation, not really finished or as adequate as needed. That is to say, these models always require further work, and importantly so. Nowhere is the difficulty in interpretability clearer than in fish recovery and its relation to habitat restoration. Referring to the aftermath of the delta smelt crisis, one CALFED ecologist asked, "Still, how will these 1000 [saved] smelt help the population recover? Is the species in recovery? As all the [ERP] restoration comes online, it will be even harder to sort out the effects."

Part of modelers' confidence in engineering models stems from the fact that in the view of many, the basic physics of the Bay–Delta system is already known but the basic biology is not. The contrast between engineering and ecological models was brought home in the CALFED gaming exercise, which relied primarily on engineering models for monthly water flows. Whereas the objective in adaptive management is to identify and focus on key uncertainties, the models in the gaming exercise were actually used to bound, at times even buffer, the gaming participants from uncertainties they all recognized; that is, they by and large treated the model runs as their firm starting point. The runs served as the participants' basis for negotiating over how to allocate water to competing uses from month to month. The effect of ending up with deficit water to cover all the requirements was translated back into the participants thinking about how they could have made decisions better and thus end up closer to a water balance. In doing so, the participants were able to identify and focus on key policy and management questions that must be answered or addressed better for future gaming exercises.

Notes

CHAPTER 1

1. These are all examples from our case studies. Of course, the literature on unintended effects of technological and environmental interventions is massive (e.g., Tenner 1996; De Villiers 2000).

CHAPTER 2

1. We thank Robert Frosch (2000, pers. comm.) for the two examples.
2. In 1995, *Science* and other journals published a Policy Forum, "Economic Growth, Carrying Capacity, and the Environment," by a group of economists and ecologists: Kenneth Arrow, Bert Bolin, Robert Costanza, Partha Dasgupta, Carl Folke, C. S. Holling, Bengt-Owe Jansson, Simon Levin, Karl Goran Maler, Charles Perring, and David Pimental (Arrow et al. 1995). The article led to a lively exchange of views involving other economists and ecologists, not only in *Science* but also in the journals *Ecological Economics* (from which we excerpted the above quote) and *Ecological Applications and Environment and Development Economics*. Our book draws heavily on this exchange.
3. The trinity of improved prices, incentives and institutions is found in the three cases on U.S. ecosystem management. For example, CAFED's water transfer program seeks to develop a policy framework for water markets based on the principle of willing buyers and sellers across different water uses, including environmental uses.
4. According to CALFED data (1999b), Delta inflow ranges from 6 to 9 million acre feet (MAF) per year, the average being around 24 MAF. The State Water Project and Central Valley Project draw an average of 5.9 MAF each year. Among the 7,000 diverters of water from the system are the water users in the Delta itself, who divert an average of approximately 1 MAF annually. These fluctuations and demands all reinforce the need to ensure water supplies are provided in a highly reliable fashion.

5. In the latest CALFED documents, the concept of adaptive management seems to be have become less prominent and more confined to the ecosystem restoration program than was the case during our interviews.

CHAPTER 3

1. This is not to say that all templates are unclear and thus inappropriate (see our proposal for settlement templates in chapter 6).
2. "In ths US, adaptive management was initially adopte din 1984 by the Northwest Power Planning Council, as a way of organizing the council's activities to protect and enhance Pacific salmon in the Columbia River Basin ... Those efforts were diverted in 1990 by litigation under the Endangered Species Act, so that the experimental phase of the Columbia basin program did not get very fair" (Lee 1999, online at www.consecol.org).
3. "In many cases, a predicate of adaptive environmental assessment and management (AEAM) has been a search for flexibility in management institutions or for resilience in the ecological system prior to structuring actions that are designed for learning. Many of the observed impediments to AEAM occur when there is little or no resilience in the ecological components (e.g., when there is fear of an ecosystem shift to an unwanted stability domain), or when there is lack of flexibility in the extant power relationships among stakeholders. In these cases, a pragmatic solution is to seek to restore resilience or flexibility rather than to pursue a course of broad-scale, active adaptive management" (Gunderson 1999a, online version).
4. Of course, many computer models are used in ecology and ecosystem management. "Computer models play diverse roles. ... They are used to design engineering structures, forecast ecosystem changes, estimate statistical parameters, summarize detailed mechanistic knowledge, and have many other applications. Such models are designed to perform well on certain narrowly defined tasks (e.g., to yield unbiased predictions with specified uncertainties for a particular process). Computer models can also be used as caricatures of reality that spark imagination, focus discussion, clarify communication, and contribute to collective understanding of problems and potential solutions. ... The role of such models is similar to the role of metaphor in narrative. The models are designed to illustrate patterns of system behavior, rather than to make specific predictions. They should be usable and understandable by diverse participants, and easily modified to accommodate unforeseen situations and new ideas" (Carpenter et al. 1999, p. 2, online).
5. Schindler (1996, pp. 18–19) argues that "[e]xpenditures on the environment typically do not increase until there is evidence of sever environmental degradation. Monies are then spent on extremely costly, time-consuming, and often ineffectual assessment, cleanup, and restoration activities."
6. The distinction between machines and ecosystems is a common one. Costanza and Geer (1995, p. 178) say about Chesapeake Bay, "But the bay is not a factory or an engine, it is an ecosystem. Instead of machinery, the bay is composed of living parts."
7. At the time of writing, details of the modeling forum could be found online at http://www.sfei.org/modelingforum
8. It is also sometimes said that other stakeholders spend more time on environmental issues than many environmentalists do. "[William Clark of Harvard University] finds that some industrial organizations, such as the Electric Power Research Institute (EPRI), an arm of the utility industry,

which produces much of the carbon emissions thorugh burning of fossil fuels, play at least as much a role in protecting the environment as do the green groups" (Shabecoff 2000, p. 30).

CHAPTER 4

1. A later section discusses service reliability sought by high reliability organizations (HROs). Weick et al. (1999, pp. 86–87) puts the matter most generally: "For a system to remain reliable, it must somehow handle unforseen situations in ways that forestall unintended consequences. This is where previous definitions of reliability are misleading. They equate reliability with a lack of variance in performance. the problem is, unvarying procedures can't handle what they didn't anticipate .... What seems to happen in HROs is that there is variation in activity, but there is stability in the cognitive processes that make sense of this activity."
2. To take a textbook example of the fallacies: because a tree provides shade does not mean that each leaf provides shade; nor does it mean that because each leaf provides little shade that the tree itself provides little shade.
3. Consider also Norton's point, "But it is odd to describe the tragedy experienced by the fishermen of Aral as resulting from 'increased uncertainty'. ... Nothing could be more certain than the decline of the [Aral Sea] fishery, as long as Soviet management ignored all danger signs for decades and transformed the sea into a desert" (Norton 1995, p. 134).
4. In the same fashion, the gradient theoretically is bidirectional, with possibilities of reducing population and extraction leading to the restoration of former ecosystem states. In reality, the gradient is sticky when moving from right to left, because reductions in population and extraction, even if they could be effected, would not necessarily be accompanied by equivalent reversals along the gradient's other three dimensions.
5. Steve Farber (1995), p. 106) asserts that ecosystem health is best maintained by the *preservation* of functions and processes that have been resilient through time.
6. A slightly more upbeat version of these asymmetries is Gunderson's paraphrase of the basic laws of thermodynamics applied to the inherent unknowability of ecosystems (Gunderson 1999b, p. 32): "(1) we can't win—we don't know enough to predict with any confidence what is going to happen in these systems; (2) we can't break even—learning by trial and error is unlikely to reveal causes and effects; and (3) we can't get out of the game—therefore, we develop some sort of system of management, based on tradition, religions, or science."
7. The link between threshold and the precautionary principle is sometimes explicit, as when Common (1995, p. 103) concludes about an Arrow et al. (1995) article, "It is implied in the article that the authors' collective judgement is that we should behave as if we are near resilience thresholds for the global system. This, I take it, is what lies behind the endorsement of the precautionary principle."
8. It should be noted that not all high reliability theorists would describe the principal features in the same way, nor would they say all such features are under the direct control of managers and decision makers. Nor, for that matter, would they maintain that the features apply to all high reliability organizations, as we shall see in a moment. Certainly, the high reliability theorists are correct in insisting that no cookbook exists to produce high reliability management in all cases that demand such

9. Performance and oversight are so important to achieving high reliability that some theorists would separate them as two distinct features of high reliability organizations (T. R. La Porte 2000, pers. comm.).
10. T. R. La Porte (2000, pers. comm.) believes that the term "culture of reliability" is general enough to embrace several of the other features of high reliability. His point serves as a reminder that, while there are characteristics shared by high reliability organizations, "there is not one clear model [of these organizations] at this time" (G. I. Rochlin 1997, pers. comm.).
11. As Gary Peterson (pers. comm. 2000) has pointed out to us, trial-and-error management may itself be difficult to realize and often rather resembles an error-and-no-trial approach.
12. Adaptive management "treats management goals and techniques as a scientist treats hypotheses. ... If expected results are obtained from experimental management, then goals and techniques are confirmed. But if expected results are not obtained, then either or both must be revised. And in the process, especially from the 'falsification' process, something new is learned ..." (Callicott et al. 1999, p. 28).
13. In other words, uncertainty, complexity, and incompleteness have different roles in each management regime. For example, consider how incompleteness is treated. In self-sustaining management, the issue is one of incomplete comprehension of the state of nature. In adaptive management, comprehension does lead to learning, but learning itself is incomplete. In case-by-case resource management, learning and comprehension take place, but in ways that do not lead to generalization. And in high reliability management, management is generalized across the system but only when information about causal processes is complete. These different forms of incompleteness, however, are connected in ways that increasingly instrumentalize incompleteness from a condition of nature outside our understanding to a property of information within our understanding.
14. Walker's extended point is worth quoting, "In the absence of detailed knowledge a rule-of-thumb policy is to maintain general ecosystem resilience through promoting diversity of species, each of which will have different responses to the environment, within different functional groups. Long-term persistence of all species in an ecosystem is best achieved by ensuring maintenance of ecosystem structure and processes (function). This, in turn calls for healthy populations of all functionally important groups (or functional types) of species. Conservation efforts are therefore most profitably spent on species which are single representatives of functional groups. Loss of such species result sin changes in ecosystem functions with cascading effects on other species. ... Species richness per se in an ecosystem has little ecological significance, but having a range of species with different environmental tolerances *within* important functional groups is a strong component of ecosystem resilience" (Walker 1995, p. 147). The Naeem viewpoint is commended in Callicott et al. (1999, p. 32): "According to Naeem (1998), 'local extinction [extirpation] within functional groups is inevitable and frequent, but reservoirs of species from adjacent ecosystems generally ensure that functional gorup or ecosystem failure, if it occurs, is likely to be transient.' Thus the maintenance of ecosystem health in humanly inhabited

and economically exploited areas depends upon the existence of proximate reservoirs of biodiversity."
15. So too are we seeing the language of engineers taking on ecosystem terminology. A senior director of information technology at Cisco Systems recently argued (Richtel 1999) that "you've got to think about having an entire ecosystem of products and services online, [with the world ahead being one] with multiple buyers, multiple customers, and multiple complementary processes."

CHAPTER 5

1. Duplication, however, does not automatically mean redundancy that enhances reliability. Conflicting and overlapping mandates within a single organization may well indicate both duplication and a decrease in reliability, as in the case of the irrigation-drainage controversy associated with the California Department of Water Resources (Hukkinen et al. 1990). What is at issue here is the nature of the decoupling.
2. Some commentators see other positive features in fragmented mandates. According to Stephen Hayward (1998), "The biggest problem with regionalism [in planning] is its premise that having major metropolitan areas divided into multiple jurisdictions is 'inefficient' and undesireable. ... University of Chicago economist Charles Tiebout posited in his 1956 article 'A level of public services that a local government should provide. Therefore, the optimal level of local public services is best determined through municipal competition, by which local jurisdictions offer different bundle sof public goods and people express their preferences by voting with their feet."

CHAPTER 6

1. Schulman (1993a) describes HROs in terms of their drive to manage fluctuations within limits and tolerances, which he calls "bandwidth management" (P. R. Schulman 2000, pers. comm.). As we shows, the drive to bandwidth management is also evident in zones of conflict.
2. We must make our own views very clear on this point. Does the argument for settlement templates legitimate letting species go extinct? No. Our argument advocates the identification of tradeoffs and setting of priorities regarding the risks of species going extinct. Under any management scenario such risks are faced, most notably the risks posed by the status quo of doing nothing more than what decision makers are now doing. What is woefully missing, in our view, is a real-world context in which decision makers can deal with these hard choices. For us that context is the settlement templates. Even a settlement template organized around the view that species extinction is unacceptable, which would undoubtedly try to move the ecosystem to the left of the gradient, poses risks to species. As one of our interviewees puts it, "all their restoration efforts necessitate taking short-term risks, which just might push some species over the edge."

CHAPTER 7

1. For more on such indicators, start with the Delft Institute's "The Ecological City," online at http://www.ct.tudelft.nl/diocdgo/.
2. Urbanization patterns different from those promoted by the current spatial planning system face the same difficult issues, including more distributed patterns giving rise to increased car mobility and transportation needs.

The evidence on the transportation implications of different urbanization patterns in the Netherlands and the degree to which these can be guided by spatial planning is, however, ambiguous (Scientific Council for Government Policy 1998; Ministry of Transportation, 2000).

3. Another noteworthy difference between our approach and the current approaches in the Netherlands is the way in which they differ iconographically. A number of publications have pointed out the importance of maps in Dutch spatial planning (Van Eeten and Roe, 1999; Beatley 2000, p. 203). Current spatial concepts to manage the Green Heart are all based in a literal sense on maps. Here what is "green" on the map is visually equated to nature reserves and "open access' agricultural areas on the ground. This reinforces the hard edges between green and non-green, where the latter are, per definition, the enemies of all that is good about the Green Heart. Our approach argues for a map in one color, green, that has different shades. In our view, greening means the recoupling of improved ecosystem functions and services across the landscape. Some areas are in need of more improvement than others. Instead of preconditioning development on a map image that is challenged every time it leaves the drawing boards of the planning agencies, development should be preconditioned on the actual improvement of the green qualities of the Green Heart and the Randstad for the people living there.

4. Much of the argument for the Netherlands holds for our other three case studies. Clearly, the greening of the Everglades Agricultural Area and Florida's coastal cities (or, for that matter, California's Central Valley and high-growth urban areas) is a priority for ecosystem management, just as is managing the Everglades Park ecosystem (or the San Francisco Bay–Delta itself). If the metropolitan water management districts involved in the Bay-Delta were to have ecosystem management responsibilities and budgets in line with those proposed for the Netherlands' National Network and in line with those that already exist for the South Florida Water Management District and Bonneville Power Administration, the United States would have in these the start of its own "national ecological network."

5. Ecological services and functions already come substantially predefined to the decision maker, so as to constrain room for creative redefinition and thus recoupling (chapter 6).

References

Alexander, E. R. 2001. Netherlands planning: The higher truth. *Journal of the American Planning Association* 67(1): 91–92.

Arrow, K., B. Bolin, R. Costanza, P. Dasgupta, C. Folke, C. S. Holling, B. O. Jansson, S. Levin, K. G. Maler, C. Perrings and D. Pimentel. 1995. Economic growth, carrying capacity, and the environment. *Ecological Economics* 15(2): 91–95.

Ashby, J. A., J. A. Beltran, M. del Pilar Guerrero and H. F. Ramos. 1995. Improving the acceptability to farmers of soil conservation practices. *Journal of Soil and Water Conservation* 51(4): 309–312.

Ausubel, J. H. 1996. The liberation of the environment. *Daedalus* 125(3): 1–17.

Ayensu, E, D. V. Claasen, M. Collins, A. Dearing, L. Fresco, M. Gadgil, H. Gitay, G. Glaser, C. L. Lohn, J. Krebs, R. Lenton, L. Lubchenco, J. A. McNeely, H. A. Mooney, P. Pinstrup-Andersen, M. Ramos, P. Raven, W. V. Reid, C. Samper, J. Sarukhan, P. Schei, J. G. Tundisi, R. T. Watson, G. H. Xu and A. H. Zakri. 1999. Ecology—international ecosystem assessment. *Science* 286(5440): 685–686.

Ayres, R. U. 1995. Economic growth: politically necessary but not environmentally friendly. *Ecological Economics* 15(2): 97–99.

Bardach, E. 1998. *Getting Agencies to Work Together: The Practice and Theory of Managerial Craftsmanship.* Washington, D.C.: Brookings Institution Press.

Barrett, G., T. Barrett and J. D. Peles. 1999. Managing agroecosystems as agrolandscapes: reconnecting agricultural and urban landscapes. In *Biodiversity in Agroecosystems*, eds. W. Collins and C. Qualset, 197–213. Boca Raton, Fla.: CRC Press.

Beatley, T. 2000. *Green Urbanism: Learning from European Cities.* Washington, D.C.: Island Press.

Beatley, T. 2001. Dutch green planning: more reality than fiction. *Journal of the American Planning Association* 67(1): 98–100.

Berry J., G. D. Brewer, J. C. Gordon and D. R. Patton. 1998. Closing the gap between ecosystem management and ecosystem research. *Policy Sciences* 31: 55–80.

Bonneville Power Administration, Army Corps of Engineers and Bureau of Reclamation. 1992. *Modeling the System: How Computers are used in Columbia River Planning.* Portland, Oreg.: Bonneville Power Administration, Army Corps of Engineers, and Bureau of Reclamation.

Brewer, J. and A. Hunter. 1989. *Multimethod Research.* Newbury Park, Calif.: Sage.

Brown, K. 2000. Ghost towns tell tales of ecological boom and bust. *Science* 290(5489): 35–37.

CALFED. 1999a. Draft programmatic environmental impact statement/environmental impact report. Sacramento, Calif.: CALFED.

CALFED. 1999b. *Facts about the Bay–Delta.* Sacramento, Calif.: CALFED. Online. Available: http://calfed.ca.gov/adobe_pdf/about_bay_delta.pdf. November 17, 2000.

CALFED. 1999c. Revised phase II report. Draft programmatic environmental impact statement/environmental impact report, technical appendix. Sacramento, Calif.: CALFED.

CALFED. 1999d. Ecosystem restoration program plan. Strategic plan for ecosystem restoration. Draft programmatic environmental impact statement/environmental impact report, technical appendix. Sacramento, Calif.: CALFED.

CALFED. 1999e. Long term levee protection plan. Revised draft. Sacramento, Calif.: CALFED.

CALFED. 1999f. Implementation plan. Draft programmatic environmental impact statement/environmental impact report, technical appendix. Sacramento, Calif.: CALFED.

CALFED. n.d. *Battle Creek Salmon & Steelhead Restoration Project,* Projects Recommended for Funding as Designated Actions in FY99. Copy obtained from the San Francisco office of The Nature Conservancy.

Callicott, J. B., L. Crowder and K. Mumford. 1999. Current normative concepts in conservation. *Conservation Biology* 13(1): 22–35.

Carpenter, S. R. 1996. Microcosm experiments have limited relevance for community and ecosystem ecology. *Ecology* 77(3): 677–680.

Carpenter, S. R., S. W. Chisholm, C. J. Krebs, D. W. Schindler and R. F. Wright. 1995. Ecosystems Experiments. *Science* 269(5222): 324–327.

Carpenter, S. R., W. Brock and P. Hanson. 1999. Ecological and social dynamics in simple models of ecosystem management. *Conservation Ecology* 3(2). Online. Available: http://www.consecol.org/vol3/iss2/art4. September 25, 2000.

Chambers, R. 1988. *Managing Canal Irrigation: Practical Analysis from South Asia.* New York: Cambridge University Press.

Clark, C. W. 1996. Operational environmental policies. *Environment and Development Economics* 1: 110–113.

Comfort, L. K. 1999. *Shared Risk: Complex Systems in Seismic Response.* New York: Pergamon.

Common, M. 1995. Economists don't read Science. *Ecological Economics* 15(2): 101–103.

Contra Costa Water District. 2000. *Mission statement.* Online. Available: http://www.ccwater.com/html/mission.html. November 19, 2000.

Cook, T. D.. R. L. Shotland and M. Marks, eds. 1985. *Postpositivists Critical Multiplism. Social Science and Social Policy.* Beverly Hills, Calif.: Sage.

Cortner, H. and M. A. Moote. 1999. *The Politics of Ecosystem Management.* Washington, D.C.: Island Press.

Cortner H. J., M. G. Wallace, S. Burke and M. A. Moote. 1998. Institutions matter: the need to address the institutional challenges of ecosystem management. *Landscape and Urban Planning* 40(1–3): 159–166.

Costanza, R. and J. Geer. 1995. The Chesapeake Bay and its watershed: a model for sustainable ecosystem management? In *Barriers and Bridges to the Renewal of Ecosystems and Institutions*, eds. L. H. Gunderson, C. S. Holling and S. S. Light, 169–213. New York: Columbia University Press.

Costanza, R., R. d'Arge, R. de Groot, S. Farber, M. Grasso, B. Hannon, K. Limburg, S. Naeem, R. V. O'Neill, J. Paruelo, R. G. Raskin, P. Sutton and M. van den Belt. 1997. The value of the world's ecosystem services and natural capital. *Nature* 387 (6630): 253–260.

Daily, G. C. 1999. Developing a scientific basis for managing Earth's life support systems. *Conservation Ecology* 3(2). Online. Available: http://www.consecol.org/vol3/iss2/art14. September 27, 2000.

Daily G. C., P. R. Ehrlich and M. Alberti. 1996. Managing Earth's life support systems: the game, the players, and getting everyone to play. *Ecological Applications* 6(1): 19–21.

DeAngelis, D. L., L. J. Gross, M. A. Huston, W. F. Wolff, D. M. Fleming, E. J. Comiskey and S. M. Sylvester. 1998. Landscape modeling for everglades ecosystem restoration. *Ecosystems* 1(1): 64–75.

Demchak, C. C. 1991. *Military Organizations, Complex Machines: Modernization in the U.S. Armed Services*. Ithaca, N.Y.: Cornell University Press.

Demchak, C. C. 1996. Tailored precision armies in fully networked battlespace: high reliability organizational dilemmas in the 'information age.' *Journal of Contingencies and Crisis Management* 4(2): 93–103.

Denzin, N. 1970. *The Research Act*. Chicago, Ill.: Aldine.

De Villiers, M. 2000. *Water: The Fate of Our Most Precious Resource*. Boston, Mass.: Houghton Mifflin.

Didion, J. 1979. *The White Album*. New York: Simon and Schuster.

Dobson, A. P., A. D. Bradshaw and A. J. M. Baker. 1997. Hopes for the future: restoration ecology and conservation biology. *Science* 277(5325): 515–522.

Duvick, D. N. 1999. How much caution in the fields? *Science* 286(5439): 418–419.

El Serafy, S. and R. Goodland. 1996. The importance of accurately measuring growth. *Environment and Development Economics* 1: 116–119.

Enserink, M. 1999a. Biological invaders sweep. *Science* 285(5435): 1834–1836.

Enserink, M. 1999b. Plan to quench the Everglades' thirst. *Science* 285(5425): 180.

Faludi, A. and A. van der Valk. 1994. *Rule and Order: Dutch Planning Doctrine in the Twentieth Century*. Dordrecht: Kluwer Academic.

Farber, S. 1995. Economic resilience and economic policy. *Ecological Economics* 15(2): 105–107.

Farrier, D. 1995. Conserving biodiversity on private land: incentives for management or compensation for lost expectations. *The Harvard Environmental Law Review* 19(2): 303–408. Online. available: http://www.lib.ttu.edu/playa/rights/r995-04.htm. September 17, 2001.

Feldman, M. S. and J. G. March. 1981. Information in organizations as signal and symbol. *Administrative Science Quarterly* 26: 171–186.

Fitzsimmons, A. 1999. *Defending Illusions: Federal Protection of Ecosystems*. Lanham, Md.: Rowman & Littlefield Publishers.

Folke, C., C. S. Holling and C. Perrings. 1996. Biological diversity, ecosystems, and the human scale. *Ecological Applications* 6(4): 1018–1024.

Fortmann, L. 1990. Locality and custom: nonaboriginal claims to customary usufructuary rights as a source of rural protest. *Journal of Rural Studies* 6(2): 195–208.

Goldstein, P. Z. 1999. Functional ecosystems and biodiversity buzzwords. *Conservation Biology* 13(2): 247–255.

Grumbine, R. E. 1994. What is ecosystem management? *Conservation Biology* 8(1): 27–38.

Grumbine, R. E. 1997. Reflections on "what is ecosystem management?" *Conservation Biology* 11(1): 41–47.

Gunderson, L. 1999a. Resilience, flexibility and adaptive management: antidotes for spurious certitude? *Conservation Ecology* 3(1). Online. Available: http://www.consecol.org/Journal/vol3/iss1/art7/index.html. September 19, 2000.

Gunderson, L. 1999b. Stepping back: assessing for understanding in complex regional systems. In *Bioregional Assessments: Science at the Crossroads of Management and Policy,* eds. K. N. Johnson, F. Swanson, M. Herring and S. Green, 27–40. Washington, D.C.: Island Press.

Gunderson, L. H., C. S. Holling and S. S. Light, eds. 1995a. *Barriers and Bridges to the Renewal of Ecosystems and Institutions.* New York: Columbia University Press.

Haney, A. and R. L. Power. 1996. Adaptive management for sound ecosystem management. *Environmental Management* 20(6): 879–886.

Harte, J. 1996. Confronting visions of a sustainable future. *Ecological Applications* 6(1): 27–29.

Hayward, S. 1998. Legends of the sprawl. *Policy Review* September/October: 26–32

Holland, J. H. 1992. *Adaptation in Natural and Artificial Systems: An Introductory Analysis with Applications to Biology.* Cambridge, Mass.: MIT Press.

Holling, C. S., ed. 1978. *Adaptive Environmental Assessment and Management.* New York: John Wiley.

Holling, C. S. 1995. What barriers? What bridges? In *Barriers and Bridges to the Renewal of Ecosystems and Institutions*, eds. L. H. Gunderson, C. S. Holling and S. S. Light, 3–34. New York: Columbia University Press.

Holling, C. S. 1996. Engineering resilience vs. ecological resilience. In *Engineering within Ecological Constraints*, ed. P.C. Schulze, 31–43. Washington, D.C.: National Academy Press.

Hukkinen, J. 1999. *Institutions in Environmental Management: Constructing Mental Models in Sustainability.* London: Routledge.

Hukkinen, J., E. M. Roe, and G. Rochlin. 1990. A salt on the land: a narrative analysis of the controversy over irrigation-related salinity and toxicity in California's San Joaquin Valley. *Policy Sciences* 23(4): 307–329.

Imperial, M. T. 1999. Institutional analysis and ecosystem-based management: the institutional analysis and development framework. *Environmental Management* 24(4): 449–465.

Johnson, B. L. 1999. Introduction to the special feature: adaptive management—scientifically sound, socially challenged? *Conservation Ecology* 3(1): Online. Available: http://www.consecol.org/vol3/iss1/art10. September 25, 2000.

Johnson, F. and K. Williams. 1999. Protocol and practice in the adaptive management of waterfowl harvests. *Conservation Ecology* 3(1): 8. Online. Available: http://www.consecol.org/vol3/iss1/art8. September 25, 2000.

Johnson, K. N., F. Swanson, M. Herring and S. Greene, eds. 1999. *Bioregional Assessments: Science at the Crossroads of Management and Policy*. Washington, D.C.: Island Press.

Katz, J. E. 1993. Science, technology and congress. *Society* 30(4): 41–50.

Kaufmann, R. K. and C. J. Cleveland. 1995. Measuring sustainability: needed—an interdisciplinary approach to an interdisciplinary concept. *Ecological Economics* 15(2): 109–112.

Keane, R., D. Long, J. Menakis, W. Hann and C. Bevins. 1996. *Simulating coarse-scale vegetation dynamics using the Columbia River Basin Succession Model—CRBSUM*. General Technical Report INT-GTR-340. U.S. Department of Agriculture, Forest Service Intermountain Research Station: Ogden, Utah.

Kloor, K. 1999. A surprising tale of life in the city. *Science* 286(5440): 663–663.

Kratochwil, F. and J. G. Ruggie. 1986. International organization: a state of the art on an art of the state. *International Organization* 40(3): 753–775.

Lackey, R. T. 1998. Seven pillars of ecosystem management. *Landscape and Urban Planning* 40: 21–30.

Landau, M. 1969. Redundancy, rationality, and the problem of duplication and overlap. *Public Administration Review* 29 (July–August): 346–358.

La Porte, T. R. 1993. 'Organization and safety in large scale technical organizations': lessons from high reliability organizations research and task force on 'Institutional trustworthiness. Paper prepared for a seminar on Man–Technology–Organization in Nuclear Power Plants, Finnish Centre for Radiation and Nuclear Safety, Technical Research Centre of Finland, Olkiluoto, Finland. June 14–15, 1993.

La Porte, T. R. 1996. High reliability organizations: unlikely, demanding and at risk. *Journal of Contingencies and Crisis Management* 4(2): 60–71.

Lee, K. N. 1993. *Compass and Gyroscope: Integrating Science and Politics for the Environment*. Washington, D.C.: Island Press.

Lee, K. N. 1999. Appraising adaptive management. *Conservation Ecology* 3(2). Online. Available: http://www.consecol.org/vol3/iss2/art3/. September 25, 2000.

Leonard, D. 1984. Disintegrating agricultural development. *Food Research Institute Studies* 19(2): 177–186.

Lerner, A. W. 1986. There is more than one way to be redundant: a comparison of alternatives for the design and use of redundancy in organizations. *Administration & Society* 18(3): 334–359.

Levy, S. 1998. *Land-use and the California Economy: Principles for Prosperity and Quality of Life*. San Francisco: Center for Continuing Study of the California Economy and Californians and the Land.

Light, S. S., L. H. Gunderson and C. S. Holling. 1995. The Everglades: evolution of management in a turbulent environment. In *Barriers and Bridges to the Renewal of Ecosystems and Institutions*, eds. L. H. Gunderson, C. S. Holling, and S. S. Light, 103–168. New York: Columbia University Press.

Lindblom, C. E. 1990. *Inquiry and Change: The Troubled Attempt to Understand and Shape Society*. New Haven: Yale University Press.

Ludwig, D. 1996. The end of the beginning. *Ecological Applications* 6(1): 16–17.

Ludwig, D., R. Hilborn and C. Walters. 1993. Uncertainty, resource exploitation, and conservation—lessons from history. *Science* 260(5104): 17, 36.

Luke, T. W. 1997. *Ecocritique: Contesting the Politics of Nature, Economy, and Culture*. Minneapolis, Minn.: University of Minnesota Press.

Mannarelli, T., K. Roberts and R. Bea. 1996. Learning how organizations mitigate risk. *Journal of Contingencies and Crisis Management* 4(2): 83–92.

March, J. G. and J. P. Olsen. 1989. *Rediscovering Institutions: The Organizational Basis of Politics.* New York: Free Press.

March, J. G., L. S. Sproul and M. Tamuz. 1991. Learning from samples of one or fewer. *Organization Science* 2(1): 1–13.

Matson P. A., W. J. Parton, A. G. Power and M. J. Swift. 1997. Agricultural intensification and ecosystem properties. *Science* 277(5325): 504–509.

Max-Neef, M. 1995. Economic growth and quality of life. *Ecological Economics* 15(2): 115–118.

McConnaha, W. E. and P. Paquet. 1996. Adaptive strategies for the management of ecosystems: the Columbia River experience. *American Fisheries Society Symposium* 16: 410–421.

McLain, R. J. and R. G. Lee. 1996. Adaptive management: Promises and pitfalls. *Environmental Management* 20(4): 437–448.

Melillo, J. M. 1998. Warm, warm on the range. *Science* 283(5399): 183–184.

Miller, A. 1999. *Environmental Problem Solving: Psychosocial Barriers to Adaptive Change.* New York: Springer.

Ministry of Agriculture, Nature Management and Fisheries. 1995. *Change and Renewal.* The Hague: Ministry of Agriculture, Nature Management and Fisheries. Online. Available: http://www.minlnv.nl/international/policy/inta/notuta1.htm. November 9, 2000.

Ministry of Housing, Spatial Planning and the Environment. 1991. *Fourth Report (EXTRA) on Spatial Planning in the Netherlands: Comprehensive Summary—On the Road to 2015.* The Hague: SDU.

Ministry of Housing, Spatial Planning and the Environment. 1996. *Het Groene Hart in woord, beeld en getal.* The Hague: SDU.

Ministry of Housing, Spatial Planning and Environment. 1998. *The Third National Environmental Policy Plan (NEPP 3): Summary.* Online. Available: http://www.netherlands-embassy.org/c_envnmp.html#nmp3. November 7, 2000.

Ministry of Transportation. 2000. *Concept eindrapport thema 2: Bereikbaarheid, bestemmingen en verbindingen.* Online. Available: http://www.minvenw.nl/rws/projects/nvvp/making_of_/themas/thema2.html. November 20, 2000.

Mintzer, M. 1995. Valuation problems and intergenerational equity issues complicate the management of global environmental risks. *Ecological Economics* 15(2): 105–107.

Moris, J. and J. Copestake. 1993. *Qualitative Enquiry for Rural Development.* London: Intermediate Technology Publications on behalf of the Overseas Development Institute.

Myers, N. 1995. Environmental unknowns. *Science* 269(5222): 358–360.

Naeem, S. 1998. Species redundancy and ecosystem reliability. *Conservation Biology* 12: 39–45.

Nathan, R. 1988. *Social Science in Government: Uses and Misuses.* New York: Basic Books.

Norton, B. 1995. Resilience and options. *Ecological Economics* 15(2): 133–136.

Ogden, J. C. 1999. Everglades–South Florida assessments. In *Bioregional Assessments: Science at the Crossroads of Management and Policy*, eds. K. N. Johnson, F. Swanson, M. Herring and S. Greene, 169–185. Washington, D.C.: Island Press.

O'Neill, R. V., J. R. Kahn, J. R. Duncan, S. Elliott, R. Efroymson, H. Cardwell and D. W. Jones. 1996. Economic growth and sustainability: a new challenge. *Ecological Applications* 6(1): 23–24.

Opschoor, J. B. 1995. Ecospace and the fall and rise of throughput intensity. *Ecological Economics* 15(2): 137–140.

Orians, G. H. 1996. Economic growth, the environment and ethics. *Ecological Applications* 6(1): 26–27.

Orr, D. 1999. The question of management. In *Unmanaged Landscapes: Voices for an Untamed Nature*, ed. B. Willers, 7–10. Washington D.C.: Island Press.

Ostrom, E., J. Burger, C. B. Field, R. B. Norgaard and D. Policansky. 1999. Revisiting the commons: local lessons, global challenges. *Science* 284(5412): 278-282.

Page, T. 1995. Harmony and pathology. *Ecological Economics* 15(2): 141–144.

Perrow, C. 1984. *Normal Accidents: Living with High-Risk Technologies*. New York: Basic Books.

Perrow, C. 1994. The limits of safety: the enhancement of a theory of accidents. *Journal of Contingencies and Crisis Management* 2(4): 212–220.

Peterson, G., C. R. Allen and C. S. Holling. 1998. Ecological resilience, biodiversity, and scale. *Ecosystems* 1, 6–18.

Pickett, S. T. A. and M. L. Cadenasso. 1995. Landscape ecology: spatial heterogeneity in ecological systems. *Science* 269(5222): 331–334.

Quigley, T. M., R. T. Graham and R. W. Haynes. 1999. Interior Columbia Basin Ecosystem Management Project: case study. In *Bioregional Assessments: Science at the Crossroads of Management and Policy*, eds. K. N. Johnson, F. Swanson, M. Herring, and S. Greene, 271–287. Washington, D.C.: Island Press.

Rhoads, B. L., D. Wilson, M. Urban and E. E. Herricks. 1999. Interaction between scientists and nonscientists in community-based watershed management: emergence of the concept of stream naturalization. *Environmental Management* 24(3): 297–308

Richardson, C. J. 1994. Ecological functions and human values in wetlands: a framework for assessing forestry impacts. *Wetlands* 14(1): 1–9.

Richtel, M. 1999. The next waves of electronic commerce. *The New York Times* December 20, C36.

Risser, P. G. 1996. Decision-makers must lead in defining some environmental science. *Ecological Applications* 6(1): 24–26.

Roberts, K. 1988. *Some Characteristics of High Reliability Organizations*. Berkeley, Calif.: School of Business Administration, University of California at Berkeley.

Rochlin, G. I. 1993. Defining "high reliability" organizations in practice: a taxonomic prologue. In *New Challenges to Understanding Organizations*, ed. K. H Roberts, 11–32. New York: Maxwell Macmillan International.

Rochlin, G. I. 1996. Reliable organizations: present research and future directions. *Journal of Contingencies and Crisis Management* 4(2): 55–59.

Roe, E. M. 1997. On rangeland carrying capacity. *Journal of Range Management*, 50(5): 467–472.

Roe, E. M. 1998. *Taking Complexity Seriously: Policy Analysis, Triangulation, and Sustainable Development*. Boston: Kluwer Academic.

Roe, E. M. 2000. Poverty, defense, and the environment: how policy optics, policy incompleteness, fastthinking.com, equivalency paradox, deliberation trap, mailbox dilemma the urban ecosystem, and the end of problem solving recast difficult policy issues. *Administration & Society* 31(6): 687–725.

Roe, E. M. and M. J. G. van Eeten. 2001. The heart of the matter: a radical proposal. *Journal of the American Planning Association* 67(1): 92–97.

Roe, E. M., L. Huntsinger and K. Labnow. 1998. High reliability pastoralism. *Journal of Arid Environments* 39(1): 39–55.

Roe, E. M., M. J. G. van Eeten and P. Gratzinger. 1999. *Threshold-based resource management: The framework, case study and application, and their implications*. Report to the Rockefeller Foundation. Berkeley: University of California.

Röling, N. G. and M. A. E. Wagemakers, eds. 1998. *Facilitating Sustainable Agriculture: Partcipatory Learning and Adaptive Management in Times of Environmental Uncertainty*. Cambridge, UK: Cambridge University Press.

Root, T. L. and S. H. Schneider. 1995. Ecology and climate: research strategy and implications. *Science* 269(5222): 334–341.

Sartori, G. 1989. Undercomprehension. *Government and Opposition* 24(4): 391–400.

Schindler, B. and K. A. Cheek. 1999. Integrating citizens in adaptive management: a propositional analysis. *Conservation Ecology* 3(1). Online. Available: http://www.consecol.org/vol3/iss1/art9. September 22, 2000.

Schindler, D. W. 1996. The environment, carrying capacity and economic growth. *Ecological Applications* 6(1): 17–19.

Schulman, P. R. 1993a. The negotiated order of organizational reliability. *Administration & Society* 25(3): 353–372.

Schulman, P. R. 1993b. The analysis of high reliability organizations: a comparative framework. In *New Challenges to Understanding Organizations*, ed. K. H Roberts, 33–54. New York: Maxwell Macmillan International.

Schulman, P. R. 1996. Heroes, organizations and high reliability. *Journal of Contingencies and Crisis Management* 4(2): 72–82.

Schulze, P. ed. 1999. *Measures of Environmental Performance and Ecosystem Condition*. Washington, D.C.: National Academy Press.

Science. 1999. Peering into 2000. *Science* 286(5548): 2240.

Scientific Council for Government Policy. 1998. *Ruimtelijke ontwikkelingspolitiek. Rapporten aan de regering 53*. Den Haag: Sdu Uitgevers.

Scoones, I. 1999. New ecology and the social sciences: what prospects for a fruitful engagement? *Annual Review of Anthropology* 28: 479–507.

Shabecoff, P. 2000. *Earth Rising: American Environmentalism in the 21st Century*. Washington, D.C.: Island Press.

Slocombe, S. D. 1998. Defining goals and criteria for ecosystem-based management. *Environmental Management* 22(4): 483–493.

Smith C. L., J. Gilden and B. S. Steel. 1998. Sailing the shoals of adaptive management: The case of salmon in the Pacific Northwest. *Environmental Management* 22(5): 671–681.

Stevens, W. K. 2000. Conservationists win battles but fear war is lost. *The New York Times* January 11, D5.

Stone, R. 1995. Taking a new look at life through a functional lens. *Science* 269(5222): 316–317.

Strange, E. M., K. D. Fausch and A. P. Covich. 1999. Sustaining ecosystem services in human-dominated watersheds: biohydrology and ecosystem processes in the South Platte River basin. *Environmental Management* 24(1): 39–64.

Tenner, E. 1996. *Why Things Bite Back: Technology and the Revenge of Unintended Consequences*. New York: Knopf.

Thomas, C. W. 1997. Public management as interagency cooperation: testing epistemic community theory at the domestic level. *Journal of Public Administration Research and Theory* 2(7): 221–246.

Thomas, C. W. 1999. Linking public agencies with community-based watershed organizations: lessons from California. *Policy Studies Journal* 27(3): 544–564.

Toman, M. 1996. Development, scale and resource valuation. *Environment and Development Economics* 1: 136–137.

United States Army Corps of Engineers and South Florida Water Management District. 1999a. *Rescuing an Endangered Ecosystem: The Plan to Restore America's Everglades.* Jacksonville/West Palm Beach, Fla.: United States Army Corps of Engineers and South Florida Water Management District.

United States Army Corps of Engineers and South Florida Water Management District. 1999b. *Central and Southern Florida Project Comprehensive Review Study: Final Integrated Feasibility Report and Programmatic Environmental Impact Statement.* Jacksonville/West Palm Beach, Fla.: United States Army Corps of Engineers and South Florida Water Management District.

United States Department of Agriculture, Forest Service; United States Department of Commerce, National Oceanic and Atmospheric Administration, National Marine Fisheries Service; United States Department of the Interior, Bureau of Land Management, the Fish and Wildlife Service, and the National Park Service; Environmental Protection Agency. 1993. *Forest Ecosystem Management: An Ecological, Economic, and Social Assessment. Report of the Forest Ecosystem Management Assessment Team ("FEMAT").* Washington, D.C.: US Government Printing Office.

United States Department of Agriculture, Forest Service, and United States Department of the Interior, Bureau of Land Management. 1994. *Final Supplemental Environmental Impact Statement on Management of Habitat for Late Successional and Old-growth Forest Related Species within the Range of the Northern Spotted Owl. Volume I (Glossary).* Portland, Oreg.: United States Department of Agriculture, Forest Service, and the United States Department of the Interior, Bureau of Land Management.

United States Department of Agriculture, Forest Service, and United States Department of the Interior, Bureau of Land Management. 1996a. *Integrated Scientific Assessment for Ecosystem Management in the Interior Columbia Basin and Portions of the Klamath and Great Basins.* General Technical. Report PNW-GTR-382. Portland, Oreg.: United States Department of Agriculture, Forest Service, and the United States Department of the Interior, Bureau of Land Management.

United States Department of Agriculture, Forest Service, and United States Department of the Interior, Bureau of Land Management, 1996b. *A Framework for Ecosystem Management in the Interior Columbia Basin and Portions of the Klamath and Great Basins.* General Technical Report PNW-GTR-374. Portland, Oreg.: United States Department of Agriculture, Forest Service, and the United States Department of the Interior, Bureau of Land Management.

United States Department of Agriculture, Forest Service, and United States Department of the Interior, Bureau of Land Management. 1997. *An Assessment of Ecosystem Components in the Interior Columbia Basin and Portions of the Klamath and Great Basins: Volume I.* General Technical Report PNW-GTR-405. Portland, Oreg.: United States Department of

Agriculture, Forest Service, and the United States Department of the Interior, Bureau of Land Management.

United States Department of Agriculture, Forest Service, and United States Department of the Interior, Bureau of Land Management. 2000a. *Summary of the Alternatives: Interior Columbia Basin Ecosystem Management Project, March 2000.* Online. Available: http://www.icbemp.gov/html/projectinfo/web/altsum.htm. November 18, 2000.

United States Department of Agriculture, Forest Service, and United States Department of the Interior, Bureau of Land Management. 2000b *The Interior Columbia Basin Supplemental Draft Environmental Impact Statement: Summary.* Portland, Oreg.: United States Department of Agriculture, Forest Service, and the United States Department of the Interior, Bureau of Land Management.

United States Department of Agriculture, Forest Service, and United States Department of the Interior, Bureau of Land Management. 2000c. *Report to the Congress on the Interior Columbia Basin Ecosystem Management Project.* Portland, Oreg.: United States Department of Agriculture, Forest Service, and the United States Department of the Interior, Bureau of Land Management.

United States Department of the Interior, Fish and Wildlife Service. 1999. *South Florida Multi-species Recovery Plan.* Atlanta, Geor.: United States Department of the Interior. Fish and Wildlife Service, Southeast Region.

United States Department of the Interior, Fish and Wildlife Service, and Department of Commerce, National Oceanic and Atmospheric Administration. 1999. *Notice of Availability of a Draft Addendum to the Final Handbook for Habitat Conservation Planning and Incidental Take Permitting Process.* Published in the *United States Federal Register*, 65(45), Tuesday, March 9, 1999: 11485–11489.

Van Eeten, M. J. G. 1997. Sprookjes in rivierenland: Beleidsverhalen over wateroverlast en dijkversterking. *Beleid en Maatschappij* 14(1): 32–43.

Van Eeten, M. J. G. and E. M. Roe. 2000. When fiction conveys truth and authority: the Netherlands Green Heart planning controversy. *Journal of the American Planning Association* 66(1): 58–76.

Vitousek P. M., H. A. Mooney, J. Lubchenco and J. M. Melillo. 1997. Human domination of Earth's ecosystems. *Science* 277(5325): 494–499.

Von Meier, A. 1999. Occupational cultures as a challenge to technological innovation. *IEEE Transactions on Engineering Management* 46(1): 101–114.

Walker, B. 1995. National, regional and local scale priorities in the economic growth versus environment trade-off. *Ecological Economics* 15(2): 145–147.

Walters, C. 1986. *Adaptive Management of Renewable Resources.* New York: Macmillan.

Walters, C. 1997. Challenges in adaptive management of riparian and coastal ecosystems. *Conservation Ecology* 1(2). Online. Available: http://www.consecol.org/vol1/iss2/art1. September 25, 2000.

Walters, C. and C. S. Holling. 1990. Large-scale management experiments and learning by doing. *Ecology* 71(6): 2060–2086.

Weick, K. E., K. M. Sutcliffe, and D. Obstfeld. 1999. Organizing for high reliability: Processes of collective mindfulness. *Research in Organizational Behavior* 21: 81–123.

Westoby, M., B. Walker, and I. Noy-Meir. 1989. Opportunistic management for rangelands not at equilibrium. *Journal of Range Management* 42(4): 266–274.

Wieners, B. 1999. "Wages of sim," An interview with Michael Schrage. *Wired* November: 110.

Wildavsky, A. 1979. *Speaking Truth to Power: The Art and Craft of Policy Analysis*. Boston, Mass.: Little, Brown and Company.

Willers, B. 1999. Toward a science of letting be. In *Unmanaged Landscapes: Voices for an Untamed Nature*, ed. B. Willers, 56–58. Washington D.C.: Island Press.

Wilson, J. A., J. M. Acheson, M. Metcalfe and P. Kleban. 1994. Chaos, complexity, and community management of fisheries. *Marine Policy* 18(4): 291–305.

Young, A. M. 1999. Invaders today, natives tomorrow? *Science* 286(5441): 901.

Index

accountability, 136, 137, 148, 178, 191, 207, 212
accounting, 152, 187, 198, 199, 226
Across Trophic Level System Simulation (ATLSS), 63, 64, 75–77, 156, 157
adaptive environmental assessment, 75, 76
adaptive management
  active, 60, 114–122, 155, 212
  passive, 114–120
aggregate behavior, 113
agricultural modernization, 13
agriculture, industrial, 3, 200
anticipation, 113
Army Corps of Engineers (ACE), 47, 64, 181
Arrow, K., 26, 53, 101, 102, 113
Ashby, J. A., 24
assessments, 45, 50, 75–78, 115, 142, 157–159
authority patterns, 108

bandwidths
  management, 8, 9, 169–218, 223, 232–236
  setting of, 164, 223

Bay–Delta Modeling Forum, 10, 75, 161
beliefs, 228
beneficial uses, 39–40, 140
beneficiaries, 191, 200
benefit–cost ratio, 212, 213
Berry, J., 22, 24, 29, 67, 85, 89, 92, 118, 134
biodiversity, 13–19, 59, 65, 89, 96–98, 119, 194–198, 222, 225, 234, 235
blind spots, 5–8, 27, 30, 67, 171–172, 181–193, 210–213, 232–238
Bonneville Power Administration (BPA), 44–46, 59–69, 73, 80, 81, 127, 129, 175, 197
Brinkhorst, L. J., 10, 218, 239
Bureau of Land Management (BLM), 41–46, 63, 66
Bureau of Reclamation (BR), 22, 43, 62–67, 142, 152, 213, 215

CALFED Program, 9, 10, 49, 65–70, 129, 137, 140, 147, 186, 209–212, 239, 241
Callicott, J. B., 16, 21, 22, 25, 57, 97, 98, 126

Case, T., 116
case-by-case management, 7, 8, 87, 95, 96–97, 104, 116–131, 168–172, 184, 205–208, 218, 228–237
Central Valley Project, 37, 41, 213, 215
Chambers, R., 131, 171, 181, 182
Cleveland, C. L., 88, 100, 103
climate change, 65, 104
Columbia River Basin, 29, 55, 63–83, 139–141, 155–158, 166, 208
co-management, 150
Comfort, L. K., 9, 123, 140, 185
Commoner, B., 101
complexity, 7, 18–27, 52–59, 64–75, 83–99, 107, 135–138, 153, 162, 169–181, 183, 192, 205, 222, 226, 231, 238
conflict, zones of. *See* zones of conflict
control room, 37, 105, 109, 127, 141, 150–159, 170, 183–187, 192, 201–204, 213, 214, 232–239
coupling, 8–19, 31, 85–99, 110–123, 130–168, 170–174, 181–229, 239
coupling–decoupling–recoupling (CDR) dynamic, 8, 9, 130–141, 170, 193, 194, 206–208
creativity, 30, 183, 194
criteria, 29, 104, 113, 118, 120, 181, 225, 231
cross-subsidy, 224
cultures, professional, 170–172, 181, 225, 231

Daily, G. C., 27, 28, 97–100, 120, 171, 222
data collection, 18
decentralization, 185
decision making, 46, 55, 78, 89, 108, 115, 162, 201, 231
decision-making units, 230
decision space, 28, 86–95, 100–119, 156, 205, 231, 237
decoupling, 8, 13–15, 110, 130–140, 146–151, 166, 170, 181, 186–199, 205–212, 221, 226–230, 236, 239

Delta islands, 36, 142, 195, 199
delta smelt, 4, 57, 63, 75, 132, 148, 153, 161, 166, 179, 189, 192, 201, 234, 241, 242
Department of Fish and Game (DFG), 22, 38, 143, 150, 185, 210, 215
Department of Water Resources (DWR), 37, 74, 80, 127, 143–145, 154, 159, 175, 185, 209–215
duplication, 136, 212

early warning, 98
ecological
 functions, 4, 13, 16, 85, 96, 97, 113, 115, 141, 194, 221–227, 235
 modeling, 64, 65, 77
ecologists, 4, 5, 9, 19–24, 31, 47, 52, 63, 72, 76, 79, 85–92, 98, 101, 104, 110, 112, 119–137, 141, 153, 154, 162, 169–171, 177–179, 184–187, 194, 201, 203, 212–214, 217, 221, 227–241
economic growth, 14, 26, 43, 101, 102, 218
economists, 26–30
ecosystem
 health, 19–23, 38, 89, 93, 100, 111–114, 140, 215, 228
 management, 13–31, 92–127
 management problems, 8, 9
 services, 3–7, 13–19, 29–31, 38, 40, 56, 68, 82–87, 91–105, 114, 116, 127–130, 141, 159, 162, 172, 180, 183, 194, 201, 227, 229, 237
Ecosystem Operations Branch (ECO-OPS), 213–215
Ecosystem Restoration Program (ERP), 40, 56, 141–147, 152, 165, 210–213, 223, 236, 241
endangered species, 3, 4, 8, 15, 22, 23, 44, 47, 55, 59–63, 73, 131–133, 141, 143, 149–151, 157, 163, 165, 176, 230–235
Endangered Species Act (ESA), 22, 55, 59, 60, 73, 132, 133, 141, 143, 149, 157, 165, 166, 175, 235
engineering, and ecology, 5, 29

engineers, 4–9, 27–31, 43–49, 58, 62–64, 85–90, 96, 103, 109, 110, 119, 120, 126–132, 138, 143, 146, 152, 154, 163, 168–171, 181, 184, 201–217, 227, 232–241
Environmental Impact Statement (EIS), 39, 50, 55, 63–67, 75, 79, 82, 96, 140, 157, 159, 177, 187, 195
Environmental Protection Agency (EPA), 21, 45, 48, 52, 161, 215
Environmental Water Account (EWA), 66, 78–83, 126, 149–152, 161, 162, 166, 186–189, 198–200, 226
error detection, 207
Everglades
  Comprehensive Plan, 64, 70–75, 81, 98, 157, 161, 163
  Agricultural Area, 47, 195, 220
evolution, 22, 30, 73, 102, 113, 117, 120, 124
experiments, 18, 22, 53–64, 80, 87, 114, 115, 137–140, 147, 155, 205, 237

failure, measures of, 121, 193
Farber, S., 104
Fish and Wildlife Service (FWS), 21, 45, 48, 56, 63–67, 79, 80, 143, 150, 215
fivefold gradient, 95, 96, 97, 228
flood control, 8, 38, 43, 61, 64, 69, 73, 74, 80, 131, 132, 142, 143, 152–156, 163, 174, 190, 195, 198, 214
Forest Service (FS), 21, 41–46, 63, 66
fragmentation, 23, 26, 67, 79, 132, 134, 223

gaming
  exercises, 151–162, 166–170, 181–186, 198–202, 214, 230, 237–240
  versus storage, 232
geographic information system (GIS), 76
goals, 8, 14–21, 27–29, 44, 52, 60, 65–68, 74–82, 96, 108, 122, 132–136, 145, 166, 174, 175, 191–195, 203, 208, 215, 218, 225, 229, 232, 236, 242
governance, 28, 104, 205
government, 21, 22, 29, 37, 48, 50, 93, 100, 111, 117, 136, 142, 200, 219–227, 239
Green Heart, 5, 10, 11, 31, 98, 129, 197–208, 218–239
Grumbine, R. E., 19–31, 57, 69, 73, 84, 86, 100, 103, 123, 126, 170
Gunderson, L. H., 9, 14, 19, 22, 32, 51–54, 60, 62, 73–79, 115, 129, 138

habitat, orientation, 61
heterogeneity, 92, 113
high reliability
  management, 5–10, 15, 24, 30, 58–60, 68–76, 82, 87, 95–130, 137–140, 168–171, 184, 203, 228–235
  organization, 24, 29, 59, 80, 105–108, 125, 159, 170, 192, 218
Holling, C. S., 22, 30, 40, 53, 62, 73, 138
housing, 14, 197–200, 218–226, 233, 240
Hukkinen, J., 9, 111, 131, 137, 162
human-dominated ecosystems, 24, 25, 56, 98, 112, 113, 195, 217, 237

Imperial, M. T., 18, 21, 29, 117, 134, 137
implementation, 6, 20, 22, 26–32, 39–55, 60, 70, 79, 81, 117–120, 132–147, 157, 189, 201, 221, 231, 234, 242
incentives, 26–30, 82, 95, 105, 107, 201, 219, 226, 227
indicators, 5, 14, 64–66, 95, 96, 102, 154, 170, 181, 186, 203, 210, 218
industrial ecology, 4, 224
information, 10–22, 45, 54, 64, 74, 77, 81, 87, 95, 100, 101, 109, 120, 123, 137, 144, 153–158, 173–179, 184, 185, 193, 199, 216, 238, 242

institutional
  innovation, 73
  reforms, 26
interagency
  cooperation, 19, 79, 81, 132–136, 166, 168
  coordination, 26, 153
  teams, 187
Interior Columbia Basin Ecosystem Management Project (ICBEMP), 41–59, 63–69, 75–81, 134, 139, 140, 155–161, 177–180, 203

Johnson, F., 9, 22, 32, 54, 65, 80, 112, 115, 129, 157

Kaufman, R. K., 88, 100, 103
keystone species, 52, 63, 75, 96, 126, 178, 179, 203, 227, 237

Lackey, R. T., 15–29, 87–103
lag time, 54, 58, 88
land-use categories, 201, 222, 234
La Porte, T. R., 6, 9, 58, 105–109, 172
learning
  modes of, 121, 228
  See also trial-and-error learning
legislation, 15, 22, 23, 31, 47, 60, 176, 220, 234, 235
levee protection, 40, 56, 122, 141–147, 213
Levee Protection Program, 40, 56, 141–147, 213
line operators, 5, 9, 28–30, 63, 109, 110, 119, 133, 144, 150–178, 184–186, 203, 213, 214, 232–242
long term, 18, 59, 66, 179, 203, 209, 210

management, real-time, 74, 147, 156, 173, 174, 201, 203, 231–236
management regimes
  characteristics of, 95, 121
  framework of, 85, 86, 104–127
master plans, 200
Metropolitan Water District of Southern California (MWD), 38, 66, 70, 191, 197–202

model-based gaming, 8, 236–238
modelers, 9, 64, 79, 159, 160, 172, 180, 232, 237–243
modeling assumptions, 155
models, 6, 18, 19, 31, 39–56, 62–66, 75–79, 86, 93, 97, 101, 104, 112–114, 120, 124, 147, 154–162, 170, 180, 181, 191, 228–231, 237, 241–243
monitoring, 19, 40, 45, 50, 55, 62–65, 73, 78, 95, 106, 115, 137, 139, 146–150, 157, 179, 210, 215, 242

Naeem, S., 125, 126
National Ecological Network, 209, 220–225, 236
National Marine Fisheries Service (NMFS), 43, 45, 63, 66, 67, 79, 80, 150
National Park Service (NPS), 21, 48, 63, 152
National Research Council, 25
natural system model, 62, 76, 156
negotiation, 22, 70, 189, 232
Netherlands, 5, 9, 11, 31, 98, 129, 197–202, 213–233
nongovernmental organizations, 21, 29, 37
normal professionalism, 171–183, 232
Northwest Power Planning Council (NPPC), 44, 46, 157

office blocks, 224, 225, 240
operational recoupling, 8, 9, 118, 120, 134, 150, 154, 160, 185–194, 207, 211–215, 221
Operations Control Office (OCO), 144, 209, 213, 214
organizational
  fragmentation, 67, 79, 132
  health, 7, 93, 110, 111, 228
  recoupling, 15
Orians, G., 60
oversight, 45, 106, 144, 166, 205, 210–215

pastoralism, 24
Perrow, C., 90, 106, 107, 173

planning, 10–18, 28–32, 38, 44–46, 50–61, 67, 72, 77–81, 141, 147–158, 174, 184–189, 198, 208, 214–233
polarization, 41, 71, 82, 119, 134, 136
policy, couplings, 191, 193
politics, 6, 74, 82, 122, 204
population
 densities, 7, 56–57, 71, 75, 85, 87, 93, 100, 104, 109, 115, 123, 130, 218, 219, 228
 growth, 6, 43, 57–60, 72, 73, 91–95, 101, 110, 116, 124, 204, 226, 230, 231
power-modeling, 75, 76, 82, 161
predictability, 7, 25, 30, 85, 86, 116
predisturbance regime, 52
preservation, 14, 15, 23–25, 73, 85, 126, 127, 178, 183, 211, 227, 229, 234, 235
presettlement template, 19, 52, 53, 59–60, 65, 75, 77, 87, 94–99, 104, 111–114, 127, 194–198, 230, 232, 240
pricing, 26, 71, 95, 200
priorities, 4, 9, 15, 20, 24, 29, 50, 68, 74, 75, 81, 89, 133–136, 157, 167, 184–203, 209–214, 225, 226, 232–235, 242
professional culture. See cultures, professional
professionals, in ecosystem management, 27–31
programming, 81, 141, 147, 231
public consultation, 80

recoupling, 8–19, 31, 85, 86, 94–99, 110, 117–123, 130–141, 147–174, 181–239
redefinition, 22, 82, 97–99, 110, 111, 118, 183, 190–200, 210, 226–235
redundancy, 70, 109, 125, 126, 135, 140, 146, 151, 166, 207
regulations, 108, 111, 143–148, 183, 184, 211, 214, 232
rehabilitation, 24, 25, 31, 44–50, 60, 71–80, 85, 96–97, 143, 177, 178, 183, 195–204, 211, 212, 226–237

replication, 64, 155
resilience, 16, 19, 23, 26, 30, 61, 65, 75, 102, 112, 113, 125, 155, 178, 196, 207
resource extraction, 6, 7, 56–62, 93, 228
resources, 5–7, 13, 15, 26, 30, 31, 37–43, 55, 61, 69, 74, 80, 82, 90–96, 101, 106–116, 122–139, 143–159, 166–170, 175, 197, 209–212, 221, 222, 228–235
responses, partial, 6, 8, 51–84, 129–168
restoration, 5, 8, 20–32, 38–40, 46–85, 96–98, 111, 115, 126–131, 137–165, 174–216, 222–229, 236, 238–243
Rhoads, B. L., 10, 24, 25, 60, 92
risk, 3, 20, 39, 41, 45, 54–61, 78, 89, 91, 99, 101, 106, 115, 119, 122, 126, 140, 144–147, 151, 157, 159, 168, 179, 183, 190, 192, 203, 211, 231
Rochlin, G. I., 10, 105–109

safety, 106–109, 122–126, 143, 145, 186, 188, 214
San Francisco Bay–Delta, 5, 6, 21, 31, 32, 109, 117, 126, 129, 139, 142, 171, 189, 209–217
scale
 large, 8, 53, 66, 139, 157, 180, 201–210, 236
 small, 9, 98, 201–210, 236
scaling up or down, 87, 205–208, 236, 237
scenario planning, 77, 78
scheduling, 59, 153, 185
Schulman, P. R., 104–110, 126, 183
self-sustaining,
 ecosystems, 95–100, 111, 120, 125–127
 management, 7, 86, 99–104, 110–113, 119, 125–127, 205, 228
settlement templates, 8, 9, 188, 198–212, 216, 229, 230, 233–238
 networking of, 204–209
short term, 59, 152, 153

South Florida Water Management District (SFWMD), 46–50, 62, 69–75, 80, 127, 195
spatial planning, 14, 198, 218–225, 233
specialization, 136, 172, 207
species-by-species approach, 22, 61
species-specific regulators, 9, 172–179, 199, 232
spreadsheets, 79, 183, 184, 239
stakeholders, 5, 27–31, 61, 69, 70, 73, 78, 80, 82, 86, 95, 120, 149, 161, 168, 180, 181, 188, 199, 211, 217, 241
  ecologists as, 29, 217, 241
State Water Project (SWP), 37, 41, 67, 74, 185, 209–215, 239
stewardship infrastructure, 100, 103, 111, 112, 126, 127, 229–234
storage, 39, 40, 49, 69–72, 81, 82, 98, 150–154, 162–168, 203, 231
strategic focus, 66
success, measures of, 121, 137
surprise, 19, 57, 58, 74, 109, 118, 151, 155, 162, 168–173, 181, 207, 215, 231, 237
sustainability, 16–21, 100, 102, 136, 140, 218
system
  boundaries, 9, 74, 188, 193–199, 210, 222, 233, 236
  interconnections, 57–59

task environment, 8, 20, 24, 59, 67, 122, 131, 137, 138, 151, 184, 205
technical competence, 106, 109, 144
technicians, 28, 154, 238
technological innovation, 73, 79
technology, 9–13, 25, 30, 43, 60, 100–107, 125, 127, 144, 174, 221, 239
terminology, 19, 25, 215, 235
The Nature Conservancy, 73, 74, 80, 91, 127, 191
Third Law of Ecology, 101
Thomas, C. W., 10, 22, 29, 30, 132
thresholds, 10, 86, 93–105, 122–124, 228
tradeoffs
  and priorities, 4, 50, 133, 184, 189–193, 199, 202, 203, 210, 212, 214, 225, 226, 232, 234, 236
  intraprogram, 184, 191
transboundary effects, 104
trial-and-error
  learning, 6, 55, 58, 108, 115, 147
  management, 114
triangulation, 118–124, 208, 209, 236, 237
turbulence, 8, 132, 137, 151, 193, 199, 207
twofold management goal, 4–6, 14, 15, 88–105, 110–121, 130–137, 183, 188, 199–202, 207, 221, 222, 227–237

unintended consequences phenomenon, 57
United States Geological Survey (USGS), 45, 48, 155
urbanization, 3, 75, 98, 195, 197, 218–226, 233

values, 14–24, 29, 30, 80, 86–93, 104, 108, 119, 146, 147, 171–180, 203, 204, 212, 218, 231, 232
Vernalis Adaptive Management Plan (VAMP), 70, 139, 186, 198, 241, 242
Von Meier, A., 159, 169–175

Walters, C., 22, 32, 40, 53, 54, 114, 138
water budget, 162–168, 188, 231
Water Conservation Area, 47, 49, 76, 163, 164
whole-system view/perspective, 66, 72–75, 97, 99, 149, 154, 178, 185, 210, 222, 224, 231
Wildavsky, A., 137, 191
Williams, K., 65
win-win, 39, 164, 189–193, 233

zero-sum game, 164
Zimbabwe, 98
zones of conflict, 6–8, 31, 71, 94, 115–123, 129–168, 169–216, 218, 221, 230–237